Robots, Zombies and Us

ALSO AVAILABLE FROM BLOOMSBURY

Advances in Experimental Philosophy of Mind,
edited by Justin Sytsma
The Bloomsbury Companion to Philosophy of Mind,
edited by James Garvey

Robots, Zombies and Us

Understanding Consciousness

ROBERT KIRK

Bloomsbury Academic
An imprint of Bloomsbury Publishing Plc

B L O O M S B U R Y
LONDON · OXFORD · NEW YORK · NEW DELHI · SYDNEY

Bloomsbury Academic
An imprint of Bloomsbury Publishing Plc

50 Bedford Square	1385 Broadway
London	New York
WC1B 3DP	NY 10018
UK	USA

www.bloomsbury.com

BLOOMSBURY and the Diana logo are trademarks of Bloomsbury Publishing Plc

First published 2017

© Robert Kirk, 2017

Robert Kirk has asserted his right under the Copyright, Designs and Patents Act, 1988, to be identified as the Author of this work.

All rights reserved. No part of this publication may be reproduced or transmitted in any form or by any means, electronic or mechanical, including photocopying, recording, or any information storage or retrieval system, without prior permission in writing from the publishers.

No responsibility for loss caused to any individual or organization acting on or refraining from action as a result of the material in this publication can be accepted by Bloomsbury or the author.

British Library Cataloguing-in-Publication Data
A catalogue record for this book is available from the British Library.

ISBN: HB: 978-1-4742-8658-9
PB: 978-1-4742-8659-6
ePDF: 978-1-4742-8661-9
ePub: 978-1-4742-8660-2

Library of Congress Cataloging-in-Publication Data
A catalog record for this book is available from the Library of Congress.

Cover design: Catherine Wood
Cover image © Bridgeman Images

Typeset by Integra Software Services Pvt. Ltd.

To Janet

Contents

Preface viii

1 Introduction 1
2 Behaviour is not enough 9
3 Machines 19
4 Intelligent robots? 27
5 Is something non-physical involved? 47
6 Zombies 63
7 What's wrong with the zombie idea? 75
8 The basic package 93
9 What's needed on top of the basic package 117
10 Functionalism 125
11 Functionalism is compulsory 139
12 Is there an explanatory gap? 151
13 Brains in vats and buckets 167

References 185
Index 187

Preface

The ancient idea of intelligent and possibly conscious machines soon gets us perplexed. Can machines really think? Could robots ever be conscious? In the first half of this book I introduce the main philosophical problems of consciousness; in the second I defend a certain approach to solving them (a controversial one, I have to say). The book is for anyone with a serious interest in these matters, not only students. No background in philosophy is necessary.

Robots are already doing work formerly done by people, and enthusiasts suggest some of them might even be conscious. Are they right? This book should help you make up your mind. But I shall not be much concerned with actual examples, which in any case would quickly be out of date. The robots discussed here are mostly imaginary, designed to illustrate fundamental philosophical points about consciousness and intelligence: points whose importance is independent of the details of particular cases. The zombies here are also a special philosophical sort: baffling – even disturbing – to think about, but not what you find in films, science fiction, or West Indian folklore (though more interesting).

There is wide agreement about what the main problems of consciousness are, but not on how to go about solving them. I argue that we are forced to take an approach according to which consciousness is a matter of the performance of *functions*. There is fierce opposition to this view. A prominent philosopher once said, 'If you are tempted to functionalism, I believe you do not need refutation, you need help' (Searle 1992: 9*). This opposition springs mainly from assumptions which, though natural and widespread, are mistaken. When we look carefully at the problems and proposed

---

* Details of books and articles are given in the References. In the text they are referred to by names and dates. For example (Searle 1992) refers to his book *The Rediscovery of the Mind*, and (Searle 1992: 9) to page 9 of that book. References to sections of chapters in the text have the form '10§6'.

solutions, and try to understand what consciousness is, there is (I shall argue) no serious alternative to functionalism – although it has to be the right kind.

The *Stanford Encyclopedia of Philosophy* (plato.stanford.edu) provides many references and much guidance on the topics discussed. There is a very accessible spoken introduction to the philosophical problems of consciousness by a leading philosopher at http://www.portrait.gov.au/stories/david-chalmers. If you want more details of the approach sketched in this book, see my *Zombies and Consciousness* ([2005] 2008).

Warm thanks to David Hankinson and Nigel Howard, non-philosophers who read and generously commented on a draft. And special thanks to my wife Janet, for all kinds of encouragement and support.

1

Introduction

1. Golden robots

Hephaistos was the Greek god of technology. Among other things, he made food trolleys with golden wheels (bling has always been with us): automata which, Homer tells us, took food and drink to the gods' meeting places and returned of their own accord (*Iliad*, book 18). Being lame, he had also constructed something even more remarkable: servants to help him get about his shiny bronze house. These robots were made of gold but 'like living young women', and able to think and speak. Two and a half thousand years ago, of course, people would have had no idea how trolleys might be made to move about independently of human agency, let alone how robots could be intelligent; these would have been seen as typical examples of wonders worked by the gods.

Still, Hephaistos was a smith and general technologist, not known for magic, and the ancient Greeks might well have supposed that what enabled him to make his robots was supreme technical skill plus an understanding of the properties of natural materials. In any case I aim to get clear about consciousness without appealing to magic – which would have amounted to giving up the search for genuine explanation and understanding.

There is a striking contrast between Hephaistos' automatic food trolleys and his golden attendants. Some of today's robots seem to be genuinely intelligent. But the golden servant-girls don't just look and behave unlike the trolleys; they are said to be conscious – and only a few misguided enthusiasts would argue that any of today's robots are really conscious. It doesn't even seem that technology has got far enough to allow us to make robots that could behave like us in all relevant ways, rather than just mimicking some aspects of human behaviour. The point, though, is that even if there are such robots, and even if some of them are genuinely intelligent, it is a further question whether they might also be conscious.

That question would not have troubled Hephaistos, since presumably all he needed was attendants who *behaved* as he wanted. That brings us to a central topic. There seems to be a difference between just behaving appropriately and really being conscious. Imagine we were faced by one of Hephaistos' attendants, and that its behaviour was indistinguishable from ours, ignoring if possible that it had metal where we have flesh and blood. Would that mean it was conscious? Would it even be truly intelligent? As will become clear, these questions not only go deep, they would continue to puzzle us after even more technological success than has already been achieved. Some people maintain that both consciousness and intelligence require no more than the right behaviour; they are behaviourists of a philosophical kind. *Psychological* behaviourists, in contrast, hold – or rather held – that behaviour provides the only basis for a truly scientific psychology: their view was influential in the first half of the last century, but has been largely superseded by the neurosciences in the last decades. Philosophical behaviourism is rejected by most philosophers today; its difficulties have come to seem overwhelming. Still, it is too important and has too many attractive features to be dismissed out of hand. Discussing it will face us with considerations that any theory of consciousness must take into account; we shall examine it in the next chapter.

A strongly contrasting view is *dualism*. Dualists are so impressed by the differences between physical (or material) things and conscious experiences that they conclude that these are two totally different kinds of things. Dualism has it that no merely physical system could be conscious; at best it would be a non-conscious husk. According to one group of dualists, we have non-physical souls or spirits; others maintain there are no such things. But all dualists agree there is something about thoughts and feelings that the purely physical component of reality cannot provide for. You don't have to be a behaviourist to oppose dualism, by the way. Many philosophers and scientists are *physicalists,* otherwise known as materialists, who hold there is nothing in the world but the physical. They are still free to reject behaviourism: see Chapter 5.

Discussing whether either philosophical behaviourism or dualism might be true will force us to get clearer about the nature of consciousness and intelligence; thinking about robots will help us to avoid focusing too narrowly on purely human mentality. If some robots are intelligent or even conscious – more to the point, if they *could* be – then their mentality might well be unlike ours in some respects; yet, for our purposes it would be enough if we could understand what it would take to make them so.

2. Conscious and unconscious perception

If some of those thoughts strike you as too remote from real life to bother about, consider the difference between conscious and unconscious perception. You're walking along, suddenly notice a cyclist riding straight at you, and dodge out of the way. What made you move? Spotting the cyclist of course – because it's by seeing, hearing, and otherwise perceiving things around us that we are able to act appropriately. Perception certainly seems to involve conscious experiences, at least most of the time. However, there is also such a thing as unconscious perception. When you were walking along just now you avoided bumping into a bollard on the pavement. Did you really see it, or did your semi-automatic avoidance result from unconscious perception? Another example: driving a car while thinking about something else, one can sometimes seem to have arrived at a certain place without consciously seeing things on the way. Is that unconscious perception? Such cases are inconclusive because they are consistent with the driver having simply forgotten the conscious experiences that nonetheless occurred. But other kinds of subliminal perception have been studied, and although there is dispute over how to decide whether there was really no conscious experience, it is hard to deny such things happen. (Evidence in one type of case turns up when a person in the experimental situation is confronted with an ambiguous word and asked to choose a particular reading. Faced subliminally with a flashed-up disambiguating word or image, they tend to favour the reading it suggests.)

The puzzling case of 'blindsight' provides another example. Some types of brain damage result in part of the person's visual field being blanked out, and they don't see – or more cautiously, claim not to see – anything in that part of the field. Yet they are still able to acquire information through their eyes about stimuli in those blind parts, as is proved by their success in answering test questions about the character of the stimuli. For example, when presented with a forced choice between alternative descriptions of the stimulus, their performance turns out to be substantially better than chance.

Non-conscious perception of any kind raises a question. What's the use of consciousness if we could get along without it? If everything we do on the basis of conscious perception could have been done equally well on the basis of non-conscious perception, what difference does consciousness make to our lives, leaving aside the important fact that experiences can be enjoyable in ways that non-conscious perception can't?

The question gains force when we consider evolution. Suppose two kinds of creatures, the Cons and the Nons, behave alike and cope equally well with their environments: they have the same capacities and dispositions. And now suppose the Cons are conscious and the Nons are not. Is it plausible that *both* should have evolved by natural selection? It seems reasonable to assume that consciousness requires properties on top of those which merely produce the right behaviour. If so, then the Cons would need more complicated internal arrangements than the Nons, so that, other things being equal, the Cons would have needed more time to evolve. In that case natural selection would have favoured the Nons, who would have had a head start. But of course we, the most successful evolved animals, are conscious, and so far as we know we have no non-conscious rivals with similar capacities and dispositions. So it is at least plausible to suggest that the capacities and dispositions provided by consciousness give some creatures an evolutionary advantage over any apparently possible non-conscious rivals. However, regardless of whether that suggestion is correct, it does nothing to answer the really baffling question. What explains the difference between the conscious creatures and the non-conscious ones? Are behavioural capacities and dispositions enough after all? And if they aren't, what else is needed? My answer is that it is a matter of the performance of certain functions. Functionalism has attracted a number of supporters over the years but is still controversial. In fact, it strikes some people as obviously wrong, if not mad (see the quotation from Searle in the Preface). I will try to persuade you that it stands out as the best approach even though it may not appeal at first sight; I think those who resist it are like the people who once argued that metal ships would never float. My defence of a version of functionalism takes up the second half of the book, starting at Chapter 8. In the first half I focus on the philosophical problems of consciousness – in particular those raised by robots and (philosophical) zombies – and on the theories that have been suggested for solving them.

3. What sort of consciousness?

Like most words that are both common and useful, 'conscious' – and for that matter 'intelligent' – has a range of more or less different meanings. It is one thing to be conscious rather than unconscious, something else to be conscious of what's happening in the street; and it is one thing to be conscious *of* something happening, something else to be conscious *that* what's happening is a birthday celebration. Self-consciousness is something else again. However, most readers will already have reasonably clear conceptions of the different varieties of consciousness, and for our purposes there is no need to catalogue and define them further. For reasons that will become clear, the main focus in this book will be on conscious *experience*. We have conscious experiences when in a normal waking state of mind we see, hear, smell, taste, or feel the things around us, for example my current experiences of the blue type on my computer screen, of the sound of birdsong, of the delightful scent of coffee. But not all conscious experiences are perceptual, although they are all subjectively more or less *like* perceptual experiences; they include hallucinations; after-images; sensations like tingles, pains, nausea, sexual pleasure; dream experiences; and the whole gamut of feelings and sensations involved in emotion. However, I believe that if we can explain how there can be *perceptual* experience, it will be easier to explain other kinds of consciousness. For that reason, the main kind of consciousness to be investigated in this book is what is involved in perceptual experience.

In a famous article, 'What Is It Like to Be a Bat?' (1974), Thomas Nagel argued that conscious experience – sometimes referred to as 'phenomenal consciousness' – poses special problems for physicalists, who maintain that our world is purely physical. He started from the common opinion that many non-human animals such as bats have conscious experiences – which I know no good reason to doubt – and went on to argue that it is impossible for us human beings, who have our own characteristic but limited set of sensory capacities, to come to know *what it is like* to have the experiences of creatures with different capacities. We might know everything about bat physiology and behaviour, Nagel argued, but that would not enable us to discover the 'subjective character' of the experiences they have by means of their system of echolocation. It therefore seems to be a big difficulty for physicalists to explain why the physical facts about us give rise to these particular kinds of conscious experience rather than to any of a whole range of different ones – or, indeed, *don't* give rise to any conscious experiences at all. How can the actual character of experience be brought within the scope of explanations based on nothing but physical and behavioural facts? How can hearing the piercing cry of a buzzard, or seeing the sun set over the sea, or smelling freshly roasted coffee, all come from what's going on in the mushy stuff inside our heads?

Such reflections led Nagel to say, 'Without consciousness the mind-body problem would be much less interesting. With consciousness it seems hopeless' (1974: 166). Jerry Fodor later went further: 'Nobody has the slightest idea how anything material could be conscious. Nobody even knows what it would be like to have the slightest idea about how anything material could be conscious. So much for the philosophy of consciousness' (1992: 5). And Colin McGinn has suggested that although there is a theory which explains how consciousness comes from processes in our brains, our minds are naturally so organized that we are constitutionally incapable of grasping it. As you may have gathered, I think all those opinions are far too pessimistic, although I do believe that explaining how there can be conscious experiences at all is the hardest and most important of the philosophical problems surrounding consciousness in general.

4. Arguments

When Nagel claims we could not come to know what it is like to be a bat, his thought is that, given that these creatures perceive the world by echolocation while we do not, there is no possibility that we should be able to learn the character of their subjective echolocatory experiences. We might be able to learn it if we somehow managed to equip ourselves with a similar perceptual system, learn to hang upside down from branches, and so on, but we are interested in what is possible given the actual facts. Which prompts me to emphasize that although philosophical views, like scientific ones, have to take empirical facts into account, they cannot usually be *tested* against facts. Typically, they have to be tested by examining supporting arguments. That usually requires us to consider what is possible or impossible, so controversy is pretty well guaranteed.

Philosophical arguments often involve thought experiments (as scientific arguments occasionally do), and their frequently outlandish character may alienate some readers. But it is worth keeping in mind that their purpose is not to offer armchair substitutes for empirical research, but to bring out the implications and limitations of common assumptions, and so to explore what is and is not possible or necessary. We have to be wary, though: although thought experiments can make valuable contributions to philosophical arguments, some are just fancy ways of reinforcing prejudices: 'intuition pumps' as Dennett (1991) calls them.

I had better add that although I have tried to be as clear as possible, even professional philosophers often need to re-read and brood over an argument met for the first time.

2

Behaviour is not enough

1. Philosophical behaviourism

Why did that man suddenly start running? Because he had stolen a handbag and noticed the police were after him. Behaviour unintelligible at first may become intelligible if we observe the person over time, or take more notice of circumstances. But we often fail to discover any clues to motives; behaviour can be inscrutable. That suggests we are right to draw a line between what people actually do and what they are thinking; so how could anyone sensibly endorse philosophical behaviourism: the view that to have thoughts, feelings, and other mental states is just a matter of the right kinds of behaviour?

Part of the answer is that behaviourists don't maintain that people's mental states will necessarily be revealed in *actual* behaviour; they also appeal to how people *would* behave in various possible circumstances: their dispositions. As I stroll round Tate Britain I may look as if I am studying and appreciating certain paintings. But that behaviour disguises my intentions. In fact, I am casing the joint with a view to devising a cunning plan to steal Turner's painting of a boat in a snowstorm. In the right circumstances I would tell you what I was up to (for example if you confided that you too were thinking of stealing that same piece of work and we decided to cooperate). Innocent though my actual behaviour appeared to be, I had a *disposition* to behave in ways that – under the right conditions – would reveal my nefarious thoughts. But given that many of our dispositions may never show up in actual behaviour, behaviourists are not being stupid when they appeal to them. With those points in mind, it is not wildly implausible to hold that having thoughts and feelings – beliefs, desires, fears, goals, and all other mental states – is just a matter of having the right behavioural dispositions. But can that really be enough?

2. Ryle on the ghost in the machine

A version of philosophical behaviourism was advocated in Gilbert Ryle's influential and highly readable book *The Concept of Mind* (1949).This attacks what he called 'the official doctrine', a widely held type of dualism originating with René Descartes (1596–1650), according to which the mind is a special kind of entity that among other things thinks, perceives, makes decisions, and possibly survives the death of the body. On this view the body is a machine, and the mind is somehow lodged inside it: a 'ghost in a machine'. Ryle was particularly concerned to show that even to think of the mind as an entity is a mistake. He argued that it is no more an entity than a car's performance is an entity (though this is not his example). The car's performance is just a matter of what it can do – its capacities and dispositions – not a special object located inside it. Similarly, having a mind is not like having a book or a brain; it is a matter of what people tend to do. We shall return to these ideas in Chapter 5. (Daniel Dennett, once a student of Ryle's, has defended a version of behaviourism more recently. See for example his 1987 and 1991.)

3. Dispositions

The idea of a disposition, then, is that even if something isn't actually doing anything, it would behave in certain ways if certain conditions prevailed. A lump of sugar has a disposition to dissolve in water – it is water-soluble – because it would dissolve if dropped in water. It has that disposition even if it never comes near any water and ends up being burnt or smashed to powder and blown away.

That example fits a simple pattern of stimulus and response. The organism, machine, or other system is so constituted that when it receives a certain stimulus, it regularly puts out a fixed response; this can be called the 'S-R' pattern. However, philosophical behaviourism is not restricted to explaining behaviour on the S-R model – not even if it is extended to include learned responses. Dispositions may be complicated in ways the S-R model doesn't fit. Consider a computer running sophisticated software, such as the sort that enables you to order goods on the internet. The firm's menu sets out a range of options, and you select one of these. Then you're offered a further range of options, and so on until you are able to select the item you want to order. In general terms, the software provides the machine with a complex system of behavioural dispositions which have the form: if x is picked, then put out menu m; if y is picked from m, then put out menu n; ... if z is selected, then perform operation o. Note particularly that this kind of software does not require the machine's dispositions to be spelt out. It doesn't require the programmer to have anticipated the customer's possible sequences of choices and to have explicitly specified each in advance. Instead, the customer's choice at a given moment temporarily fixes the machine's disposition: the choice gives the machine a disposition to acquire certain particular further dispositions from then on – this in turn being a disposition it didn't have until that moment – and so on. In addition, the software gives the machine dispositions to alter certain other dispositions if certain conditions are satisfied, together with further dispositions to alter those, and so on.

These systems of dispositions result in a suitably programmed computer's behaviour being very complex; impossible to pin down in a few words, and very far from just a set of built-in stimuli and responses. Now, Descartes maintained that no machine could accurately mimic human behaviour. The fact that computers and sophisticated programming didn't exist in the seventeenth century meant that he knew of no examples of machines behaving in ways that would lead us to think they had thoughts or feelings. There were indeed ingenious automata driven by clockwork, but in spite of performing very elaborate sequences of behaviour, all their actions were built in by their makers. Today, given what we know about the capacities of computers, it starts to seem possible that a suitably programmed system might produce behaviour essentially like ours; on which see the next chapter.

4. More about behaviourism

You might agree that the notion of behavioural dispositions is very flexible, yet suspect that even so, mental states cannot be just a matter of having the right dispositions. Consider toothache. Why did I wince just now? Because I felt a sudden sharp pain in a tooth: that sensation *caused* my disposition to wince. But a cause can't be the same thing as its effect: a thing can't cause itself. So it seems the sensation is not the same thing as the disposition. Isn't that an objection to philosophical behaviourism?

It does seem like a serious difficulty. Before investigating it, let's consider mental states other than consciousness, for example *believing* and *wanting*. If I believe there is gold in those hills, I'm disposed to agree if an old timer remarks, 'There's gold in them thar hills.' If I'm properly equipped and want to get rich, I myself might become disposed to go to the hills and start digging. At the same time I might want to conceal my belief from other people. In that case, although in public I might contradict the old timer's remark – perhaps disguising my trip as a camping holiday – I should retain a disposition to agree with him in circumstances where others wouldn't notice I was doing so – maybe recording it in a private diary.

Typically, I acquire beliefs on the basis of knocking about in the world and seeing, hearing, and otherwise perceiving things. According to the view that beliefs are dispositions, acquiring beliefs consists of *nothing but* acquiring certain kinds of dispositions, which can stand in complex relations among themselves and to other kinds of mental states. They can include dispositions to alter certain other dispositions if certain conditions are satisfied, and further dispositions to alter those, and so on – as with sophisticated computer software. Nor does behaviourism about beliefs and wants require all beliefs to have been acquired via perception; it permits them to be acquired as a result of interactions among the dispositions that constitute other beliefs, as when detectives work out from what they already know that the murderer must be X.

Here is a further weighty consideration about believing, wanting, hoping, and other mental states that don't seem to necessarily involve conscious experiences. It is at least plausible, as we have seen, that they are just a matter of having certain dispositions. But if they are not, what else could they be? Internal states, certainly; informational states, no doubt. But dispositions are internal states, and the ways they are caused by our perceptual encounters with things in the world and by interactions among themselves suggest they can also be informational states. So there is a lot to be said for philosophical behaviourism when it comes to mental states not essentially tied up with consciousness. Consciousness itself, on the other hand, remains a problem for the theory. Let us consider a couple of apparent counter-examples.

5. The perfect actor

Suppose I decide to behave just as if I were suffering from toothache when I'm not. That means I am disposed to behave as if I had toothache. Since behaviourism entails that being in pain actually *consists* of having such a disposition, doesn't this example demolish the theory?

No. Imagine I stick to my resolution to behave as if I had toothache even when I'm offered huge financial rewards for coming clean, or threatened with terrible penalties if I don't (I'd have to be a fanatic, but perhaps that's what I am). Even so, I'd still be left with other dispositions connected to the very fact that my present behaviour results from a decision rather than a toothache. Among the information I still have is that I am in fact following that decision, and according to behaviourism my possession of that further information is itself a matter of having certain dispositions: ones that I should not have had if I had genuinely been suffering from toothache. So although I might be a perfect actor, the very fact that I am pretending to feel pain when I don't ensures that I don't have the right dispositions. The objection fails.

But what if I came to forget why I had started out on this perverse project in the first place? I would end up no longer having the dispositions which at present show I am only pretending to feel pain. Such selective amnesia would leave me with nothing but the dispositions of someone really in pain. You might think that vindicates the objection – but it doesn't. If I end up with all and only the dispositions of someone really in pain, then according to the behaviourists I really am in pain – when the perfect actor objection was supposed to provide an example of having the right dispositions without really being in pain. If the objector insisted that nevertheless my amnesia would leave me pain-free, they would only be contradicting their behaviourist opponents – begging the question – not describing a counter-example.

So I think the perfect actor fails as a counter-example to philosophical behaviourism. However, there are more promising cases.

6. The Giant

Let's be clear to start with that if philosophical behaviourism is true, then *anything* with the right package of behavioural dispositions has mental states, regardless of whether it looks like one of us, or is alive, or has evolved, or was put together by Frankenstein. Otherwise the behaviourist story would have to be supplemented by information about the structure of the system in question, its evolutionary history, and so on, contrary to the whole point of behaviourism. It might be a natural organism, but it might just as well (if possible) be an artefact, or a pantomime horse or, indeed, the Giant.

This is an enormous model of a human being controlled by a team of ordinary people inside it. They receive information on screens and dials, and control the thing's behaviour by pressing buttons and moving levers like an aircrew. These controllers not only make it behave like a giant human being; their commitment (for they too might be fanatics) ensures that the Giant has the dispositions of a human being.

But does the Giant really have thoughts and feelings? We can see it doesn't. The only thoughts and feelings around are those of its controllers, and they are not at all the same as the thoughts and feelings the Giant would have, if it had any. When the Giant says, 'I'm getting tired, I'll sit on this rock and take a rest', that is only because its controllers reckon that a giant with real thoughts and feelings would be feeling tired by this time, and so they cause its speech machinery to produce those words, and its limbs to make it sit down. But there is nothing that could constitute feelings of tiredness in the Giant itself, nothing that could constitute its working out what to say, and nothing that could constitute its deciding to utter those words. In fact the Giant is just a sophisticated puppet, whose puppeteers happen to be inside rather than outside. It is no more a subject of mental states than a glove puppet or a pantomine horse. If philosophical behaviourism were true, though, the Giant would really have thoughts and feelings, since it has the dispositions one would expect a real giant to have. If it doesn't really have thoughts and feelings, philosophical behaviourism is false. Or so it seems. (The principle which applies here is that if we know 'If P, then Q', then we can deduce by strict logic that 'If not Q, then not P' – though not, of course, that 'If not P, then not Q'.)

That is a pretty convincing counter-example on the face of it, certainly more convincing than the perfect actor. But here is a possible reply. Behaviourists could object that the behaviour described is in a sense not the system's own because it is determined by someone other than the supposed subject. It all results from decisions by its controllers, just as with puppets – and no one could claim that puppets were counter-examples to behaviourism. It is not unreasonable to insist that in order for behaviour to count it must be originated by the system itself, not by something distinct from it, and especially not by human controllers. If software equips a computer or robot with the capacity to work out its own behaviour on the basis of its own information, then that gives us a sense in which it is autonomous. So, skirting for the moment the questions raised by the idea of its 'own' behaviour and information, the Giant is not after all a wholly persuasive counter-example to philosophical behaviourism.

You may be hard-headedly sceptical about the value of far-fetched imaginary cases. But thought experiments can help to deepen our grasp of the relevant concepts and their interrelations, and so to work out the logical implications of theories. Of course any attempted solutions to philosophical problems must be consistent with the facts, and especially with scientific knowledge – science being the best way we have of discovering what the natural world is like and how things work. But philosophical problems don't usually arise from factual ignorance: they typically reflect a failure to *understand* aspects of the world, perhaps because we are gripped by mistaken assumptions. Discussing imaginary cases is one way to discover whether our assumptions are consistent. If your theory leads to a contradiction it can't be true; and thought experiments often help to expose contradictions.

7. Externalism

While the objections to philosophical behaviourism discussed in this chapter are not conclusive, they point to real difficulties. Underlying these difficulties is the pretty well irresistible thought that actually having a sensation such as pain is something actually happening, not just a matter of dispositions however complex. Such reflections underlie the widely held opinion that philosophical behaviourism is moribund if not dead. But the insights that made it attractive have led to interesting developments. Ryle emphasized that the mind is not an object; instead we should think of mental states as constituted by our behavioural dispositions – which typically involve interacting with the world around us. My belief that it's raining involves dispositions such as taking an umbrella if I go outside (I'll do something else if I actually want to get wet). Obviously, plenty of other dispositions are comparably linked to that belief; in fact it seems each belief is connected to a whole complex network of dispositions. So a behaviouristic account of believing seems quite plausible; and it would surely also apply to wanting, hoping, fearing, and many other mental states.

A natural question: what ensures that my belief really is *that it's raining* and not something else? Without the influence of behaviourism it was easy to assume that what gave beliefs their contents and words their meanings were facts about what was going on in believers' heads, independent of links with the outside world: internal facts about people's minds. That assumption made it easy for Descartes to envisage the sceptical possibility that there might be no external world such as we suppose there is, with other bodies, the sea, sun and moon, and all the rest – but that a malicious demon had arranged for him to have the same experiences and other mental states as he would have had if such an external world had really existed. If you are influenced by behaviourism, you will not be too ready to accept Descartes' suggestion because you will not assume that the contents of our thoughts depend entirely on the contents of our heads. Instead, you will tend to say that what gives beliefs their contents is the ways in which they are related to features of the environment by complex networks of behavioural dispositions. The view that the contents of mental states and the meanings of words are not determined by what goes on in the subject's head but depend on their relations to the outside world is *externalism*. Hilary Putnam famously summed it up: 'Cut the pie any way you like, "meanings" just ain't in the *head*' (1975b: 227). Many philosophers agree.

8. Conclusion

Like the perpetual motion machine, behaviourism would have been marvellous if it had worked. It still has a lot going for it in connection with beliefs, desires, and other so-called 'intentional' states, but it seems hopeless when applied to sensations, emotions, and consciousness generally. ('Intentional' and 'intentionality', by the way, are the generally accepted labels for a much wider range of states than *what is intended*. Their distinctive feature is that they are about something or have content.) As preparation for discussing the idea that a suitably programmed robot might be a counter-example to behaviourism, the next two chapters consider machines generally, and in particular (without getting technical) computers and robots.

3

Machines

1. Computers

For centuries the only machines were things like windmills, watermills, and clocks: powered by wind, water, weights, springs, or sometimes by animals or people. More recently, steam power became available, and then electrical power. And in the last century we learned to use electronics and computers. One question is whether the differences between mechanical and electronic systems matter from the point of view of our interest in intelligence and consciousness. Here we can take advantage of ideas developed by Alan Turing, whose pioneering analysis started from a consideration of mathematical computation. He made clear how a huge range of different activities could be represented by the performance of a few very simple basic operations – no matter how the machines actually performing those operations might be powered.

The aim of this elementary and, I hope, user-friendly section is to describe the fundamental features of any computer, features present even in the first electronic computers which, as everyone knows, took up a vast amount of space but had tiny memories compared with those of today. From our point of view size doesn't matter. What does matter is the basic features shared by all computers from the most primitive to the most sophisticated. We get a painless introduction to these features by considering an ordinary diary. It is divided into a sequence of spaces or *locations*, each corresponding to a day of the year, in which information can be stored: these make the diary as a whole a sort of *memory* system. You can *write* information into any of these locations; you can *read* whatever information they contain; you can *copy* stored information into a different location; you can *delete* it. The diary's memory locations may contain information in the form of data (such as 'Lucy's birthday') or instructions (such as 'Buy bread').

A computer too has a large number of memory locations into which data or instructions can be written and stored, and from which such information can be read, deleted, or copied. However, a mere diary won't actually carry out any of the instructions you may have written in its memory locations. It just waits passively for you to do the work yourself, and for you to write or delete data or instructions. The distinctive feature of any computer, in contrast, is that it will do whatever you tell it – provided you tell it nicely, in language it understands. Equipping a diary with its own ability to execute instructions would transform it into a computer.

To make an ordinary diary execute instructions would take magic, but the computer works in essentially simple and non-magical ways. To start with, as most people know, all the information put into it is normally encoded in binary form: as sequences of '0's and '1's. For example, the Arabic numerals '1', '2', '3', '4', and so on are represented in the binary system by '1', '10', '11', '110', and so on (where '10' doesn't mean *ten* but the result of adding 1 to 1 (that is, *two*); '11' encodes not *eleven* but the result of adding 1 to binary '10' (i.e. *three*), and so on). Letters and other characters are also encoded by sequences of these binary digits '0' and '1', so that, for example, upper-case 'A' might be encoded as '00000001' and lower-case 'a' as '00000010'. A further feature is that each memory location has its own 'address', also encoded as a number.

Two things are vital for understanding how the computer follows its instructions, that is, how it runs a *program* (which is just a sequence of instructions). One is that the instructions composing the program are broken down into a handful of extremely simple basic types of operations (such as copying the contents of one memory location to another; adding the number stored at one memory location to the one stored at another and putting the resulting sum into yet another). The second is that each of those few basic types of operations is performed by a special device, typically an electronic circuit. For example, there is an *addition* device which, when two sequences of '0's and '1's are put into it, causes their sum to be stored at a specified address in the machine's memory. Whatever any computer does, no matter how sophisticated the software it is running, is done by a handful of such devices performing those basic operations in accordance with the programmed instructions.

When we use computers we are not normally aware that these basic operations are being performed. This is partly because we don't normally do the programming ourselves: we don't have to key in any instructions, except in the greatly simplified form of clicking buttons on a menu. To do our own programming – as of course the first computer scientists had to – would have required us to work out how the tasks we want the machine to perform could be done by means of the few basic operations: not usually very easy. Once we had managed to work out what the relevant basic operations must be, and in what sequence they must be performed, we should then have to list these instructions (write the program) in terms that would make the machine work. The program would then be put into the machine's memory. The most basic type of program uses the computer's own 'machine code'. This is normally expressed by means of sequences of binary digits. And here is a vital point: when such a program has been stored in its memory and activated, it will *cause* the machine to perform the listed operations. The stored sequences of '0's and '1's consist of arrays of electronic components each in the appropriate one of two possible states (one state for '0', the other for '1') and when these components are made to transmit signals through the machine's electrical connections, they activate the circuits which actually perform the basic operations mentioned above, such as adding two numbers.

The computer won't do anything at all, apart from things like humming and reflecting light, unless it's running a program. The program might simply make it print the letter 'a' until you stop the machine. On the other hand it might make it produce valuable medical diagnoses when data about a patient are put in. Whether the machine acts stupidly or intelligently depends entirely on what program it is running.

Here we can take a quick glance at artificial intelligence (AI) : the long-running, highly developed, and in many ways highly successful project of devising computer programs to make systems do things that would require intelligence if done by human beings. (Evidently, there cannot be a sharp dividing line between AI and robotics.) If behaviourism is correct, then, given AI's success, we already have artificially intelligent systems. But there is no need for workers in AI and robotics to maintain that their systems are genuinely intelligent. Intelligent behaviour is enough for their purposes, as it was for Hephaistos. However, the following question might occur to you: Does the computer – the machine itself – understand the instructions it is following? You may think that if it can follow them, it must understand them. And if it understands them, doesn't that mean it is really intelligent – just because it is following them, and even if they make it do stupid things? I wouldn't go so far. Consider what is actually happening. One lot of physical objects (the circuits which physically represent the '0's, '1's', and other components of the instructions) cause changes in another lot of physical objects (the circuits which represent locations in the machine's memory). These changes, when we interpret them in the ways intended, constitute the execution of the instructions. Does that justify saying the machine itself understands its program? To see why I think that is doubtful, compare a piano. It is so constructed that when the key for middle C is struck, processes occur which result in middle C being sounded. It would be strange to suggest the piano understood the significance of that key being struck. What is certainly true of the computer is that, like the piano, it is so constructed that when acted on in a particular way – specifically, by the occurrence of the physical events which constitute the workings of its program – it is caused to go through a sequence of processes which have a certain result intended by its programmers. But failing any further considerations, we seem to have about as much reason to ascribe intelligence to the computer itself as we have to ascribe intelligence to the piano. When we come later to consider what is needed for something to be genuinely intelligent, we shall see that the computer itself, considered in isolation from any program, doesn't have what it takes.

2. More about machines

What I have been saying about the workings of a computer is rather abstract. It emphasizes the functions performed, not the physical details of whatever might be used to perform those functions. This is because those details are irrelevant when the question is whether or not something is a computer. Electronic systems may be fastest at present, but although speed is obviously important, the point here is simply to get clear about what, quite generally, a computer *is*. Charles Babbage's prototypes would have been caused to run by complicated arrangements of gears powered by human muscle; and steam or water pressure or clockwork could perfectly well be alternative sources of energy for computers. When it comes to memory, there is a whole range of storage devices built from a wide range of different materials. But from the point of view of what it takes for something to be a computer in the first place, all that matters is that it performs the right functions.

We classify computers as machines. However, they are not what people in earlier centuries had in mind when they discussed the relation between ourselves and machines. It will be useful to define 'machine' so as to cover both computers and many other systems. For our purposes we need only consider a subclass of the type of system known as a 'Turing machine' (in honour of his groundbreaking work). It is an 'abstract' system in the sense that the definition does not specify any physical details.

We can get a grip of the main ideas with the example of a very simple slot machine. It offers either a chocolate bar costing £1, a packet of crisps costing £2, or both. You can press any of three buttons marked 'Bar', 'Crisps', and 'Confirm', one at a time but in any order. If you want just a chocolate bar or just a packet of crisps, you press the appropriate button and then the 'Confirm' button. The machine then displays the amount to be paid and delivers the goods when it receives the money. If you want both products, you press both buttons, one after the other, and then 'Confirm' as before. The essentials of this machine's activity can be represented abstractly in terms of *possible inputs, possible outputs*, and *possible internal states*. The inputs are pressings of the three buttons; the outputs (for our simplifying purposes) are the machine's displaying 'Pay £1' or 'Pay £2' as appropriate, together with its delivering the goods on receipt of payment.

The vital point is that the machine needs some distinct *internal* states too, to ensure it can register the fact that a button has been pressed without confirmation. If you press 'Bar', for example, the machine needs to be able to retain the information that you have done so in order to be able to react appropriately to your going on to press 'Confirm' or 'Crisps'. (To keep things simple, you must start from scratch with this machine if you want more than one of either of its products, so that if you press the same button twice it displays 'Press Confirm'.) The fact that this machine can retain information is tremendously important. It means it can store information and therefore has a kind of memory.

The machine's operations, including the facts about its memory capacity, can be abstractly represented by the following table, where its three possible internal states are represented by '0' (for 'initial state'), 'B' (for '"Bar" has been pressed'), and 'C' (for '"Crisps" has been pressed'). (If this is starting to make you feel faint, skip to the next section.)

Examples of what each line means: *top line*: if 'Bar' is pressed when the machine is in its initial state, it moves to state B and waits for further input. *Fifth line*: if 'Crisps' is pressed when the machine is in state B, it reverts to its initial state, displays the amount to be paid, and delivers the goods on receipt of payment.

We can generalize from this example for later reference. There is a type of Turing machine which has a finite number of possible inputs $(I_1, I_2 ..., I_n)$, a finite number of possible outputs $(O_1, O_2 ..., O_m)$, a finite number of possible internal states $(S_1, S_2 ..., S_l)$, and a *machine table* that specifies, for each possible internal state and each possible input,

Input (press)	Current state	Next state	Output
'Bar'	0	B	No action
'Crisps'	0	C	No action
'Confirm'	0	0	No action
'Bar'	B	B	'Press Confirm'
'Crisps'	B	0	'Pay £2'; delivers goods when paid
'Confirm'	B	0	'Pay £1' etc.
'Bar'	C	0	'Pay £2' etc.
'Crisps'	C	C	'Press Confirm'
'Confirm'	C	0	'Pay £1' etc.

which state and output come next. (Strictly, the 'machine' is just the abstract specification, independently of how it might be instantiated.) One concrete instance of an abstract machine of that kind is a computer running a program. If the program was at all complicated, the machine table would have to be very long to take account of the enormous amounts of information storable in the machine's memory. In our example there are only three internal states: 0, B, and C, so the memory can store only a tiny amount of information. To represent the information in a typical computer's memory, in contrast, would require many millions of different internal states. Yet from our point of view that is beside the point. What matters is that the workings of any computer could (in principle if not in practice) be represented by means of a machine table of the sort just illustrated.

Very large numbers of what we should ordinarily classify as machines are of that kind ('finite deterministic automata'). These provide a useful basis for moving on.

3. Only machines?

Imagine that Hephaistos had anticipated recent developments in technology, and that his golden attendants were in fact computer-controlled robots. What, if anything, would that imply for the claim that these robots could not genuinely think or understand language because they were 'only machines'? We know that their controlling computers – or strictly, the programs they were running – might be tremendously complicated: even more complicated than the most sophisticated in use today. Two striking examples – although both are being superseded by others as I write – are Watson, which produces (often correct) answers to general knowledge questions in a natural language such as English, and in 2011 consistently beat two star competitors in the American quiz show Jeopardy; and Amelia, who uses more than twenty languages, learns from documents, and solves problems on the basis of the information she acquires. Many people would be inclined to keep emphasizing that those robots would still only be machines. But can anyone say it is obvious that a machine controlled by a computer could not have thoughts or feelings? After all, it once seemed obvious that the sun was about the same size as the moon and that the earth was flat. The claim that robots could not be intelligent, or that they could not be conscious, needs to be defended by argument, not prejudice.

4. Levels of description

Like other kinds of machinery, whether evolved or artificial, computers can be described from different points of view and at different levels. Because I was aiming at a broad understanding of fundamental principles, I described the main features of computers in terms of functions such as storing information, adding numbers stored in different locations in memory, and so on. But our interest in computers might have arisen from other points of view, for example from that of a designer who already has a specification in terms of the functions to be performed, and needs to work out how best to implement it in terms of physical components: which configuration of components, and which particular components, will best fit that specification? Or we might have been interested in the hardware components themselves, comparing ones made in accordance with different designs, or from different materials. Evidently, these components in their turn can be described at different levels.

More relevantly, there are also different levels of functional description. Suitably programmed computers will perform such high-level functions as diagnosing medical conditions; translating between languages; predicting the movements of rockets designed to land on comets – with increasing success as the relevant kinds of software are developed. In earlier centuries only human beings could have done such things, so it is natural to describe the behaviour of computers or computer-controlled robots in terms that imply something like human intelligence. 'Diagnosing', 'translating', and 'predicting' seem to have that implication. But they also imply that the machines do such things as think about the information they are given, consider different possibilities, assess probabilities, and much else – activities which all imply intelligence. Are we misusing the words when we describe the operations of these machines in such terms, or are we literally correct? The next chapter contains arguments that will be useful for dealing with such questions.

4
Intelligent robots?

1. Robots?

There seems to be no generally agreed definition of 'robot'. But there are no generally agreed definitions of 'intelligent' or 'conscious' either, and that doesn't stop us from discussing philosophical questions about intelligence and consciousness. We come to these discussions already equipped with at least rough and ready conceptions of what the key words mean, and there is much to be said for the view that it is only after going into the implications of such conceptions that we can assess the value of any proposed definitions – by which stage we might no longer need them. All the same, I will say briefly how I am using the word 'robot'.

I apply it to what you have when a suitably programmed computer is in control of a system with something like a sense receptor (perhaps a computer keyboard, perhaps light-sensitive cells with suitable connections to the controlling program), and a capacity for interacting with things outside it (a computer screen, a voice, an arm). The computer itself includes an essential component of any robot worth considering: a system for storing and processing information. As for that tricky word 'information', we can take it to stand here for the contents of the controlling computer's memory locations. More abstractly, it stands for those of the system's internal states that are typically caused by its sense receptors' exposure to its environment, and typically contribute to guiding its behaviour (see Chapter 8).

Very strong claims are made for what robots could eventually be made to do. Homer said Hephaistos' golden attendants were intelligent (they had *nous*), and he implied they were conscious. Evidently, Hephaistos' knowledge of robotics was far ahead of ours – amazingly wide and deep though the latter is. No doubt he knew things we don't. At any rate we have not yet reached Hephaistos' technical level. Perhaps, though, there are fundamental obstacles to the idea of intelligent artefacts, hence, for reasons to be discussed later, to conscious ones. In any case, technical feasibility is not what matters from a philosophical point of view. Our question must be whether there are any reasons for ruling out such robots altogether.

2. Misguided worries

A common objection to the idea of genuine intelligence in computer-controlled systems is that they only do what they have been programmed to do. They can't think or reason for themselves, it is said, and so, like the Giant, have no mental states of their own and are completely unintelligent. But this objection only reflects ignorance or prejudice: it doesn't explain why there couldn't be a program to endow a system with human-like reasoning abilities. Some people assume the robot's responses to questions or other inputs must have been anticipated by the programmer, a fixed response being assigned to each input; they assume it must be an S-R system (see 2§3). That is a mistake. Typically the software ensures that the system constructs its own response, based on its classification of the input. This is true even of relatively simple machines. Even a pocket calculator isn't an S-R system: it doesn't incorporate a table beginning 'input 1+1, output 2; input 1+2, output 3 ...'. Instead, it has a circuit which causes it to produce the sum of whatever numbers are keyed in.

A different objection: computers and other machines are made of the wrong kind of stuff to be capable of genuine thinking and reasoning. Doesn't that require real neurones, neurotransmitters, and so forth? But why should these make a relevant difference? Suppose we discovered creatures on another planet whose behaviour and capacities were consistent with their being genuinely intelligent. And now suppose we found their brains were made from entirely different materials from ours. Would that force us to conclude they were not intelligent? Why should ours be the only possible physical basis for intelligence? We have good evidence that other materials could do at least some of the necessary work. Various kinds of man-made prosthetics provide working substitutes for faulty pathways in the nervous system. For example, cochlear implants use electronics to detect and encode sound and then stimulate the auditory nerve to allow deaf individuals to hear.

Another objection: surely, robots couldn't have free will! I have two comments. First, I know of no good reason why robots shouldn't have free will – at least in the sense in which we ourselves have free will. Of course these matters are controversial, but my second comment explains why that need not affect our investigations: there are no plausible arguments showing that free will is essential for intelligence or consciousness! Plenty of people – at any rate plenty of philosophers – rightly or wrongly claim they themselves lack free will, which presumably they would not be able to do if they were not both conscious and intelligent; and it would take better arguments than I know of to show they are mistaken. You may object that the real point is that robots are usually deterministic systems; their behaviour is determined by their construction and inputs. True, some bits of machinery exploit random processes such as radioactive decay, and are not deterministic like normal computers; but that doesn't affect the main points. The trouble with the objection is that many people hold that we ourselves are deterministic systems. In any case, the endlessly debated question of whether determinism is compatible with free action doesn't seem to have any necessary connection with the questions that concern us. If there is an objection to robotic intelligence or consciousness hereabouts, I have yet to learn what it is.

An objection some people raise is that a robot has no soul. But what is a soul? Perhaps they mean the robot lacks something special which we ourselves have – something non-physical and not capable of being reproduced in the materials that go to make up the robot. But what is the soul supposed to do? Is it essential for understanding and thinking? To assume that it is, hardly amounts to an objection to the view that future robots might have thoughts and feelings. The objector seems just to take for granted that only non-physical beings could think or feel. That would imply it matters what a thinker is made of – which exposes this objection to the same reply as an earlier one: why shouldn't souls be made of copper and silicon? (I wouldn't absolutely rule out the possibility that there might have been non-physical beings with thoughts and feelings.)

So much for some relatively feeble objections; more cogent ones will be examined shortly. Notice, by the way, that in these discussions we can legitimately disregard those aspects of behaviour which depend on our specifically human nature, for example on distinctive features of our anatomy.

3. The Turing test

Alan Turing thought the question 'Can machines think?' was not clear enough to be properly investigated, and of course he had a point. In a famous article he proposed an alternative: given a suitably programmed computer, could an interrogator tell whether it was a machine or a human being – just by putting questions to it over a limited period? He predicted that 'in about fifty years' time' it would be possible to program computers so that 'an average interrogator will not have more than 70 per cent chance of making the right identification after five minutes of questioning' (Turing 1950: 57). In the past decades there have been claims about programs which allegedly pass this test, but for reasons that will become clear they need not concern us. Instead, we can concentrate on certain philosophical examples and arguments, starting with a highly influential one due to John Searle: the Chinese Room argument. This targets not only the Turing test, but also the view that a suitably programmed computer would have any cognitive states at all (beliefs, desires, intentions, and the like). Searle is attacking what he calls 'strong AI', according to which 'the appropriately programmed computer literally has cognitive states' (1980: 353). But his own denial of this thesis is also a strong claim. (By the way, the Turing test just described has nothing to do with Turing machines. Most Turing machines – that is, concrete machines whose abstract specification qualified them for that description – would fail the Turing test.)

4. The Chinese Room

Searle exploits the fact that computer programs consist of instructions that could in principle be followed by a human being. Of course computer programs (software) with any degree of sophistication are so large and complicated that a human being working at normal speeds would not get very far with the task of actually carrying out the instructions (given that computers perform millions of basic operations per second). But even if that leads you to reject Searle's argument, it is still extremely interesting. In any case the question is theoretical: is the Turing test any good as a test of intelligence? Searle argues that it isn't, and that is an important claim.

To begin with we have to imagine that there is a program which endows a computer with the ability to understand Chinese. So far as I know there is no such program, but again the point is theoretical. There can be no objection to Searle's starting from the assumption that there is one, since his argument is a 'reductio ad absurdum'. This means that if sound, it will show there couldn't be such a program. In outline, the argument goes: 'If there is such a program, then it will enable the machine genuinely to understand Chinese. We can see that in fact it will not do so. Therefore there could not be such a program.'

Imagine now that this program has been converted from computer language into English and conveniently stored in Searle's laptop (we can forget that no such things existed in 1980, when his argument was originally published). He sets up this laptop in a closed room with a slot in the door through which Chinese speakers can post questions to him in Chinese characters. The key steps of the argument are these. On the one hand Searle himself doesn't understand Chinese; the characters on the pieces of paper posted into his room are for him mere meaningless squiggles. On the other hand he can operate the allegedly Chinese-understanding program which instructs him how to produce replies (also in Chinese characters) to the messages that come in. So on receiving one of these sequences of characters, he looks up the English instructions, which will be on the following lines: 'Check through the sequence of characters in the message. If the first is [squiggle-squiggle], then go to instruction 10,123 in the program and execute it using characters [squaggle-squaggle] and [squoggle squoggle]. If the second character is ... [and so on at great length and with enormous complications]'. The final instruction is to put out whatever sequence of characters results from having followed the instructions. Keep in mind now that the program's definition ensures that Searle's responses will be, in context, just like those which might have been put out by a Chinese speaker.

Searle argues that if the program really endowed a computer with an understanding of Chinese, then, since he is doing everything the program would have made the computer do, running the same program ought to have endowed him with an understanding of Chinese. But clearly it doesn't. He no more understands Chinese after working with the program than he did at the start. To him the characters remain meaningless squiggles. So his reasoning is as follows: he starts off not understanding Chinese; the program is supposed to make the computer understand Chinese; he does what the computer does; he still doesn't understand Chinese; therefore, the program doesn't make the computer understand Chinese.

Notice that if this ingenious and much discussed argument works, then it shows more than that the Turing test is not a good test of intelligence. It also shows that no program at all, regardless of how sophisticated it might be, could endow a system with understanding, or with beliefs, desires, or similar states. But *does* it work? Even today, plenty of people think it does.

5. The Systems Reply

The most powerful objection, I think, is what Searle calls the Systems Reply. It appeals to the fact that a computer's behaviour at a given time depends entirely on what program it is running. If the program just makes it print 'Goo' regardless of which keys you press, then the machine will look pretty silly. If on the other hand it is running a really good chess-playing program, then it might make you look silly. The point is that the computer itself, considered independently of any program it may be running, is not a candidate for intelligence at all. (As we noticed in the previous chapter, it might possibly be said to understand its program; but that is the kind of understanding a piano has.) The serious candidate for intelligence is not the computer considered on its own, but the system: the entity which consists of a machine running a suitable program.

When those considerations are brought to bear on the Chinese Room argument, it becomes clear that Searle has made a large assumption. He assumes that the fact that he doesn't understand Chinese entails that the system consisting of him running the special program doesn't understand Chinese either. But why should his opponents concede that? Their view is that a suitably programmed system would understand Chinese. They can therefore reply that whether or not he himself understands Chinese is irrelevant. It is irrelevant just as it is irrelevant whether or not one of his neurones understands Chinese. His opponents are happy to agree that the computer by itself, considered apart from any program it may be running, doesn't understand Chinese. Since Searle's role in the system is comparable to that of the computer, it is no objection that he doesn't understand Chinese.

Searle has a reply. He says that the individual in the room can simply memorize the program. (Considering the vast size and complexity of any program that could plausibly be claimed to endow a system with an intelligent grasp of language, it seems highly unlikely that even Searle himself could perform this amazing feat of memory. But yet again we must keep in mind that the point is theoretical. He can legitimately assume that the trick *could* be done by an individual with superior cognitive powers.) The result, Searle says, is that

> The individual then incorporates the entire system. ... All the same, he understands nothing of the Chinese, and a fortiori neither does the system, because there isn't anything in the system that isn't in him. If he doesn't understand, then there is no way the system could understand because the system is just a part of him. (1980: 359)

You may find that convincing. But Searle's opponents can reply that he has simply missed the point. It doesn't matter whether the program is stored in the individual's own memory, or in a huge library of reference works, or in his laptop. What matters is that the operator remains a different entity from the program he or she has memorized. The operator is one system; but the operator and the program together make a different system. You might still find that idea hard to swallow; I will try to make it palatable.

Obviously the program must ensure that the system's replies to the Chinese sentences put into it make overall sense. If the replies betrayed more than the normal human kinds of inconsistency and other sorts of incoherence, the system would be unmasked: Turing's interrogator would judge it not to be intelligent. So the program needs to be designed to make it seem that the interrogator is dealing with a normal human speaker of Chinese. And this means that – a point not to be missed – the Chinese speaker represented by the program must have a reasonably coherent system of beliefs and wants, with no more than a normal human amount of inconsistency. Imagine, then, that Searle, or whoever the operator may be, dislikes the works of Hegel and loves the paintings of Jackson Pollock. But imagine also that the program represents the fictional Chinese speaker as an admirer of Hegel and a hater of Pollock's work. The program makes the system sometimes produce Chinese sentences expressing those attitudes, such as, 'Hegel is the greatest philosopher of all time', and 'Pollock's stuff is rubbish.' Now, those sentences don't express the opinions of the operator, who of course still doesn't understand the Chinese sentences. So whose opinions do they express? All we can justifiably say is that they express the views of the fictional subject imagined by the programmers. And that subject is plainly not the same as the operator, whoever the operator may be. Therefore Searle's suggestion that the operator could memorize the program makes no relevant difference to the situation. Whether inside the room or memorized inside the operator's head, the program doesn't become 'just a part' of the operator. The system made up of the operator running the program is still distinct from the operator. So the Systems Reply is untouched by Searle's reply to it.

You might suspect that Searle must surely acquire some knowledge of Chinese in the course of operating the program. But why? All he has to do is manipulate what are, to him, just squiggles. And remember that the instructions are in English, not Chinese. As Searle tells the story, he is the operator in the room, and knows the point of the exercise. But he might just as well have told the story about someone else who was not only ignorant of Chinese, but didn't even know that the squiggles were meaningful in any language. Indeed, they might actually *be* totally meaningless so far as this person knew; the operator could follow the instructions without knowing that the squiggles were meaningful. Nor, of course, do the instructions make any connections between strings of characters and things or situations in the world: there is nothing on the lines of 'Use this string of characters if you want to express gratitude.'

6. The Robot Reply

If the Systems Reply demolishes Searle's Chinese Room argument, that doesn't mean his thesis is wrong. It doesn't mean it was a mistake to deny that an appropriately programmed computer could literally have cognitive states. He may be right about that even if his argument is defective. A very interesting (and often overlooked) fact about his original paper is that he has a *different* argument for that denial. This second argument consists of the claim that 'symbol manipulation by itself couldn't be sufficient for understanding Chinese in any literal sense' (1980: 359f, see also 362). And while I think the Systems Reply demolishes the Chinese Room argument, I agree with this last claim: a different and much stronger argument. This is because, like Searle himself and many other people, I don't see how expressions could be meaningful if they weren't somehow or other connected to things in the world. One key fact about wants, beliefs, and related states is that they are *about* things: they have intentionality. Now, the Chinese characters featuring in the program are connected to things in the world – not meaningless squiggles – because they belong to a human language whose speakers use its expressions to interact with one another, exploiting and generating multifarious connections. Yet the computer knows nothing of these connections for the same reason that the human being operating the imagined program knows nothing of them, or might as well know nothing of them. The Chinese characters might be meaningless squiggles so far as the operator is concerned, and the same goes for the computer. The Systems Reply is to the effect that it is simply irrelevant if neither operator nor computer taken by themselves understand Chinese, since what Searle's opponents claim is that the system made up of the operator working a suitable program would understand Chinese. But Searle's recently quoted remark – which amounts to a second argument – encapsulates a powerful objection to this view. It implies that no such system could understand Chinese because there are no appropriate connections between the expressions in its inputs and outputs, and the rest of the world. So it doesn't really have intentional states.

The 'Robot Reply' offered by some of Searle's opponents is intended to meet this objection. This reply is that even if a computer on its own could not understand Chinese, a computer-controlled robot could be programmed to behave more or less like one of us, and to produce utterances that were related to other people and things in much the same ways as the utterances of Chinese speakers. A robot whose computer was programmed in that way *could* understand a language (the Robot Reply goes) because its utterances would have the right sorts of connections to things in the world. So its states, in contrast to those of a computer on its own, would have genuine intentionality.

Searle's reply to this objection is to modify his argument so that it applies to a program of the kind envisaged: this program is supposed to control a robot, not a computer sitting on its own. He takes an English version of this program into the room and sets to work playing the part of the robot's controlling computer. What would have been inputs to that computer are communicated to him in his room by radio, and as before he follows the instructions and puts out messages which are transmitted to the robot and, among other things, cause the robot's limbs to move just as its internal computer would have caused them to move. But, Searle says, he still doesn't understand what's going on. 'I don't understand anything except the rules for symbol manipulation', he says, and so 'the robot has no intentional states at all' (1980: 363).

I suggested in the last section that Searle doesn't properly appreciate the Systems Reply, he doesn't get it. I think that same failure shows up again in his reply to the Robot Reply. He puts all the weight on the point that *he* doesn't understand Chinese and has no idea how his activities are connected to a robot. But just as exponents of the Systems Reply can legitimately point out that it is irrelevant that he doesn't understand Chinese, so exponents of the Robot Reply can point out that it is irrelevant that he doesn't understand interactions between the robot and the rest of the world. It remains true that the whole system – the robot controlled by the program – interacts with things around it, and not only produces Chinese sentences appropriate to the utterances produced in its hearing, but also says things in Chinese that are appropriate to its circumstances. So I don't see how Searle's reply to the Robot Reply can succeed.

Still, as before, that doesn't mean his claim is mistaken. Showing that his arguments don't work doesn't mean his conclusion is false. Now let us recall the Turing test. It seems that Searle's reasoning fails to settle whether the Turing test is a good way to establish whether something is intelligent. However, some time ago Ned Block devised an interesting argument to do that job. As with the imaginary cases discussed earlier, considering it will help to deepen our grasp of what it takes for there to be genuine thinking and feeling.

7. Block's machines

Turing's proposal was to replace the unclear question 'Can machines think?' with 'Can a machine pass this test?' Recall that the test is for an interrogator to try to make the system reveal itself, within a fixed time limit, as a machine rather than a human being, purely on the basis of what is keyed into it and what it puts out. Block doesn't think Searle's Chinese Room argument is a satisfactory refutation of the claim that the test tests intelligence, and proposes something strikingly different. A useful preliminary will be to consider two kinds of chess-playing programs.

One is the highly sophisticated kind actually used in the project of beating human players. (A well-known example is 'Deep Blue', which in 1997 beat the world chess champion. It won two games out of six, the champion won one, and there were three draws.) Although these systems use enormous memories, they also have features which can be said to enable them to weigh up the advantages and disadvantages of possible future moves: features which could be said to add up to a degree of intelligence. The other kind of program uses 'brute force', exploiting the fact that in chess there is only a finite number of possible positions (ungraspably many, though: something like 10^{120}. Producing a program using brute force in this way would be practically impossible, but we are still concerned with theory, not with what is practical). From each of these possible positions, excluding those which end the game, there is a good next move (it doesn't have to be the best). The program would list each possible position and assign a good next move to it. When it was the computer's turn to move, it would scan the board to establish what the current position was and put out the move assigned to it. In contrast to the first kind of program, this kind surely has no claim to make the system intelligent, or to have any mental states at all. It has no more intelligence than a piano. Pianos also produce a fixed output for a given input, but we don't credit them with intelligence – otherwise it might seem a good idea to give them piano lessons.

Those two examples certainly represent a contrast, but it doesn't follow that the distinction between genuine and fake intelligence is sharp. Consideration of other possible systems will show, I think, that in plenty of cases our everyday notion of intelligence doesn't enable us to say definitely whether or not they are intelligent. The present point is only that brute force systems cannot be counted as intelligent according to our everyday notion. (Should we count programs such as Watson and Emilia as unintelligent just because they depend on mountains of stored information taken from encyclopaedias, dictionaries, and so on? I leave you to ponder this question!)

Block starts his argument by defining a revised version of the Turing test in terms of 'conversational' intelligence. What he calls the 'Neo-Turing conception' of conversational intelligence is

> the capacity to produce a sensible sequence of verbal responses to a sequence of verbal stimuli, whatever they may be. (Block 1981: 18)

This might not seem to get us very far. Isn't 'sensible' too close to 'intelligent' to be useful? But Block is not defending the Neo-Turing conception; he is attacking it. The point is that he needs a fair statement of his opponents' position, and he might as well have used 'intelligent' rather than 'sensible'. For his claim will be that even if the verbal responses would count as sensible or intelligent, the fact that a system can produce them doesn't guarantee that it is intelligent. He goes on to describe a way of programming a computer so that it will satisfy the Neo-Turing conception of intelligence in spite of not being intelligent. In some respects the idea is comparable to that for the brute force chess program.

We start by fixing a time limit, which may as well be Turing's five minutes. Then we discover the maximum number of characters typable in five minutes; pretend it is 1,000. Next, we cause all sequences of up to 1,000 characters to be printed out. (The program would be very simple, but the number of sequences it would generate would be truly monstrous. Assuming there are sixty different characters, the number of all sequences of up to a thousand characters will be 60^{1000}: 60 multiplied by itself a thousand times. Pause a moment to reflect on what will be included: colossal quantities of rubbish of course, but also, for example, true five-minute biographies of everyone who will ever have lived. Plus astronomically many false biographies. Plus powerful philosophical arguments not yet thought up, together with plausible but defective ones; and indeed absolutely everything capable of being written in five minutes, whether staggeringly important or deadly dull.) Most of these sequences of characters will be nonsensical, but among the rubbish will be the ones we need to pick out. These are sequences that can reasonably be construed as conversations between two people where the second participant's contributions are sensible in the context of the whole conversation up to that point. (Don't ask how we pick out these possible conversations from the masses of rubbish: we might persuade planet-fulls of research students to do it.)

All these possible five-minute conversations are stored in the computer's memory, marking off the segments purportedly contributed by each participant. The computer is then programmed as follows. When a sequence of characters is keyed in, the computer searches the stored possible conversations for the first one whose initial segment matches what has been typed in so far. It then puts out the next segment of that same possible conversation. And so on.

The interrogator is free to key in any sequence whatever, including nonsense. All that matters is that the second participant's contributions (to be thought of as the system's own responses) make sense in the context of the whole conversation. The sequences need not all be in English, so long as they are constructed from characters on a normal keyboard and – as with the Chinese Room program –

the second participant is represented as a normally coherent human being. Here is one possible exchange:

> *Tester:* 'Do you know any French?'
> *Computer:* 'No – could you teach me some?'
> *T:* '"I'm hungry" in French is "J'ai faim". "J'ai" means "I have", and "J'ai faim" is literally "I have hunger". Now, "un chat" means "a cat". So what do you think "J'ai un chat" means?'
> – and so on.

A computer running this program will clearly have capacities which ensure that it satisfies the Neo-Turing conception of intelligence. Regardless of what ingenious or bizarre material the interrogator challenges it with, its responses will all be sensible in the context – provided the time limit is observed. This limit is essential because without it there would be indefinitely many possible inputs and the programmers could not even in principle anticipate all of them. Notice that if the interrogator tries to trick the system by putting in the same sequence more than once, it is unlikely to produce the same response each time; probably it will give different responses. In this respect the system stands in contrast to the brute force chess program, which assigns the same move every time the system is faced with the same position on the chessboard. That is emphatically not how the Block machine works. Its response to a given input depends not only on that input itself, but also on the whole history of the 'conversation' up to that input's arrival. Since the machine is guaranteed to respond in ways we should count as sensible in context, it is unlikely to give the same output to a given input within the five-minute test period. Of course it may repeat its performance in *successive* tests, but that is irrelevant: the Neo-Turing conception takes account only of what it would do for a single five-minute period – that is all it needs to do.

You might object that the system will be ignorant of what is going on around it. That too is beside the point, since what is being tested is intelligence, not knowledge, whether general or specific. It need have no sources of information beyond its program and whatever the interrogator puts into it. And don't forget that the programmers will have anticipated some likely tests: Suppose the tester keys in, 'What do you think about these terrible storms we've been having?'

A stonewalling reply might be, 'I just don't know what to think – I've really not been able to attend to current affairs.' No matter what the tester keys in, the programmers will have given the machine a reply that is intelligent in context, even if it makes the machine seem ignorant or heartless.

So the Block machine satisfies the Neo-Turing conception of conversational intelligence. But Block maintains we know enough about it to be sure it has 'the intelligence of a toaster' (1981: 21). He remarks that 'all the intelligence it exhibits is that of its programmers' (which is true but misleading, since – as he points out – it could theoretically have come into existence by chance, perhaps as the freak result of an electric storm in the computer centre, involving no intelligence at all).

8. The realism of everyday psychology

Behaviourists are committed to the view that the machine Block describes really is intelligent. But if they continue to maintain that view, they are ignoring something we have already noted, and which pretty well everyone I have asked agrees on: that it is at any rate a *necessary* condition of intelligence according to our everyday notion that the system in question works out its own response to what it sees as its situation. If that is right, the everyday or 'folk' psychology we use in everyday life – our ordinary scheme of explanation for our own and others' behaviour in terms of wants, beliefs, and feelings – is not purely behaviouristic. It includes a moderate degree of realism about the nature of our internal processes, and doesn't allow wants, beliefs, and the rest to consist purely of behavioural dispositions. Of course, you may object that the folk-psychological scheme isn't scientific and needs to be replaced. The trouble is that no one seems to have worked out a usable alternative. Anyway, my reason for agreeing with Block that his machine would not be intelligent is that in accordance with the moderate realism of folk psychology, genuine intelligence requires the system to be capable of working out its own behaviour on the basis of its own assessment of its situation. (Although I think this condition is necessary, the examples of Watson and Emilia suggest it might well not also be sufficient.) If that is correct, the Block machine is not intelligent because all its responses have been programmed into it from the start. It doesn't even understand what it says: it would produce total rubbish as readily as good sense. Nor, of course, can it work out syntactic or semantic analyses of the interrogator's inputs or its own outputs as we ourselves do, if only unconsciously.

Although the machine engages only in verbal behaviour, Block explains how to apply the same ideas to other kinds of behaviour too. Instead of possible ten-minute conversations, this machine would be programmed with possible total life histories of a given number of years (say up to 100). On reasonable assumptions this kind of robot would produce intelligent behaviour in any possible situation. But we need not consider the details: I take it Block has shown that a system could in principle have all the right behavioural dispositions without being genuinely intelligent. If that is right, the Turing Test of intelligence is inadequate, and we have yet another reason to reject philosophical behaviourism. The conclusion is that when considering whether an organism or artefact is intelligent, we have to take account not only of how it behaves and would behave, but also of what goes on inside it. (Is there anything useful to be said about these internal processes? See Chapters 8 and 9.)

9. An intriguing argument

Some philosophers are fond of appealing to logical or mathematical theorems, hoping their crystalline solidity will lend their own reasoning respectability. One of the most remarkable mathematical results of the last century has been used against mechanism – the idea that we are machines – and has therefore been thought to provide ammunition against physicalism. If that is right, robots cannot possibly be genuinely intelligent. The reasoning is interesting, though it may not be to your taste.

The mathematician and logician Kurt Gödel (1906–78) explored the long-pursued project of doing for elementary arithmetic what Euclid's axioms did for geometry. Starting from those axioms it is possible to prove a large number of geometrical theorems. It seemed reasonable to adopt this approach to arithmetic: look for a compact set of axioms that would generate all the truths of arithmetic – but only truths: no falsehoods. A set of such axioms, associated with logical rules of inference for proving theorems from the axioms, constitutes a *formal system*.

Gödel's *incompleteness theorem* undermined that project. He proved that if a formal system for elementary arithmetic is consistent (proves nothing but truths), then it cannot also be complete (cannot prove all the truths). In his proof he showed that for any such system there is a mathematical statement which competent logicians can discover to be true, but is not provable in the system.

The philosophical argument based on that theorem starts from the assumption that if mechanism is true, then each of us can be regarded as the 'concrete instantiation' of a formal system: my beliefs can be regarded as the formal system's axioms, and my patterns of reasoning can be regarded as its rules of inference. If for example I happen to believe the moon is made of green cheese, then 'The moon is made of green cheese' is one of the axioms of my formal system. Another might be 'the moon is bigger than Wales', and from those axioms my usual patterns of reasoning will enable me to derive the conclusion 'There is a piece of cheese bigger than Wales.' This will be a theorem provable in my formal system.

The assumption, then, is that if mechanism is true, then we all instantiate formal systems whose axioms represent our beliefs; shortly I shall offer an objection to this assumption. The argument starts from the reasonable premiss that competent logicians all know elementary arithmetic. Given the assumption just noted, it follows that if competent logicians are machines, then the formal systems they instantiate must include axioms for elementary arithmetic. But Gödel's theorem applies to all formal systems which include the axioms for elementary arithmetic, assuming those systems are consistent. Consider, then, the formal system supposedly instantiated by a certain competent logician Alf. By Gödel's theorem, there is a sentence of that formal system which he can discover to be true, but is not provable in the system. But if Alf really were a concrete instantiation of that formal system, then, by the theorem, he would *not* be able to discover the truth of that sentence – when we have just seen that he *can*. So, given Gödel's theorem, the hypothesis that Alf instantiates a formal system which includes axioms for elementary arithmetic leads to a contradiction: that he both can and cannot discover the truth of the sentence in question. It follows that that hypothesis is false. Alf cannot be an instantiation of such a formal system, and is therefore not a machine after all. And that proves that not *all* human beings are machines, in which case mechanism (according to which we *are* all machines) is false. Therefore, on reasonable assumptions, physicalism too is false.

This philosophical argument stirred up a lot of discussion (see Lucas 1961). Among the objections to it is one attacking the assumption of consistency. Surely, even the best logicians harbour some inconsistent beliefs? If they do, the argument collapses. For if the formal system supposedly instantiated by Alf were inconsistent, then he could prove the arithmetical sentence in question because *any statement whatever* follows logically from a contradiction. A different objection, which I find much more interesting, depends on points made in the last chapter about levels of description and explanation. It focuses on the argument's key assumption that if a person is a machine and instantiates a formal system, then the system's axioms correspond to the person's beliefs. When you think about it, it is hard to see how that assumption could be true.

For we noticed that the behaviour of computers and, for that matter, of computer-controlled robots, can be described at different levels. At high levels we talk of them diagnosing medical conditions or playing chess; at low levels we talk of them deleting and replacing data in memory locations, retrieving and adding numbers from different locations, and so on. Now, the software engineers who design programs for computers work at a number of levels, using different 'high-level languages'. But once the software has been decided on and installed in a computer, what actually does the work is a version of the program at the lowest level: that of the 'machine language'. At this level the whole program is encoded by long sequences of '0's and '1's – and it is these (typically in the form of electronic circuits) which actually cause the machine to perform the required operations. If the system as a whole is playing chess, for example, 'playing chess' is a high-level redescription of sequences of behaviour brought about by processes describable in the machine language, so there is no more to the activities which qualify for the high-level redescription than huge numbers of very simple low-level activities. And of course that goes also for the behaviour which counts as 'doing arithmetic'. Suppose we have a very sophisticated robot, Fla, whose behaviour resembles that of the mathematician Alf, and that Fla is adding the two numbers 1234 and 5678 to produce the sum 6912. Fla might do this 'in its head', it might use a calculator, or it might use pencil and paper. But 'adding two numbers in its head', 'using pencil and paper to work out the sum', 'using a calculator', are all high-level descriptions of Fla's activities. And – the key point – those descriptions do not figure in the machine-table version of the robot's program. We can say the robot as a whole knows such things as the rules of elementary arithmetic and has such and such beliefs because its program ensures that when it is performing low-level operations, such as adding or subtracting numbers, its behaviour and dispositions qualify for those high-level descriptions. But that doesn't mean the rules of elementary arithmetic are explicitly included in its program – even though the programmers, when designing the program, may have stated them in some high-level language. So it doesn't mean that the system's *beliefs* correspond to any statements occurring in the machine-language version of its program.

A computer running a given program can be viewed as instantiating a Turing machine, its program being by accepted principles equivalent to a formal system composed of axioms and rules of inference. But if what I have been saying is on the right lines – unfortunately it would take too long to go into detail – it is a mistake to assume that the formal system instantiated by someone who can do arithmetic must itself include axioms for arithmetic. In that case the attempt to refute mechanism and physicalism by means of Gödel's theorem is misconceived.

10. Conclusion

We have seen (I think) that some robots with human-like behavioural dispositions are not genuinely intelligent. If that is right, intelligence needs more than the right behavioural dispositions, and we can finally dismiss philosophical behaviourism. But clearly, we are not entitled to conclude that no possible robot could be intelligent. Indeed, I know of no serious arguments to that effect other than the ones already rejected. I think there are good reasons to suppose that robots could be intelligent – and indeed it is hard to resist the claim that Watson and Amelia (or, to be pedantic, robots programmed like them) are intelligent. Perhaps robots might become more intelligent than us, as some fans claim will happen quite soon. (The extreme vagueness of everyday notions of intelligence still permits judgements of a suitably vague sort.) So far, then, we can't rule out Homer's intelligent trolleys or even his conscious golden servant girls. But to make further progress we shall need more consideration of both intelligence and consciousness. To prepare for that, we must take note of a range of views about the nature of consciousness.

5

Is something non-physical involved?

1. Cartesian dualism

If any kind of dualism is true, thinking and feeling involve something non-physical, which means that no robot could be intelligent or conscious. Physicalism, on the other hand, according to which everything is physical, leaves plenty of scope for conscious and intelligent robots – though physicalism has its own difficulties, as we shall see. Dualism and physicalism are the most widely held views on the nature of consciousness, though not the only ones.

Descartes' dualism was particularly clear-cut. According to him there were just two kinds of stuff or 'substance': minds and bodies. Each of us consists of a mind coupled with a body. Minds (souls, spirits) are distinguished from everything else by the fact that they think – and thinking includes pretty well every kind of mental activity. The mind is 'a substance whose whole essence or nature consists in thinking' (Descartes 1641, *Meditation* 6). Bodies, in contrast, are distinguished by the fact that they occupy space. Each of these two kinds of substance has effects on the other. When we see, hear, smell, or otherwise perceive things, these bodily events have effects on our thinking. Equally, thinking often causes changes in our bodies and so, indirectly, in the rest of the world, as when we decide to go for a walk. So the physical world of things extended in space causes changes in our minds, and our minds cause changes in the physical world; hence the name 'interactionist' dualism.

Descartes thought our bodies were mere machines. However, not all of our behaviour can be explained by the workings of these machines. To make his points vivid he invites us to try to imagine a machine that would look and behave like a human being, and goes on to argue that such a thing is impossible. No machine could behave like one of us, he thought, because it would not have reason; in particular, it would not be capable of originality or creativity. Sometimes when we find ourselves in a familiar situation, we don't need to work out what to do, we act automatically. In such cases our Cartesian minds make no contribution. But very often we do have to work out how to act, and in these situations, Descartes held, our reason determines what we do in a way that mere machinery could not. Language provides striking examples. We can both construct and understand any of indefinitely many different sentences, while so far as Descartes knew no machine could do that. The best a machine can do, he assumed, is produce a range of fixed responses to possible inputs. Similarly for non-verbal behaviour. Unlike us, the machine could only produce stereotyped responses, rather than different ways of behaving tailored to suit arbitrarily various situations. If I were suddenly to lose my mind, my bodily machinery might run on for a while – heart beating, lungs pumping air; I might even walk or sing in a vacant sort of way. But without a mind my behaviour could not show distinctively human features. So no machine could even behave like a human being. Animals, on the other hand, in Descartes' view, do not have minds at all. They are machines pure and simple: automata whose behaviour could all be explained in terms of mechanical interactions among their parts.

The trouble with this reasoning – we can say with hindsight – is that the seventeenth century knew nothing of computers. That is why Descartes could say that in order to produce anything like human behaviour, the machine would need to have 'a particular disposition for each particular action' (*Discourse* 5). That is, it would be an S-R system: it would have to be so constructed that for each of a given set of possible stimuli, it would put out a particular fixed response. Today we know that even relatively simple machines can produce the right response to a given input without listing each possible input in advance. As we noticed earlier, a pocket calculator doesn't incorporate a table beginning 'input 1+1, output 2; input 1+2, output 3 ...'. The reason it can produce the sum of two numbers is not that its designer worked out all possible sums and listed them, but because it has circuits which make it produce what are in fact the right results; in a sense they make it work out those results. That example shows what is wrong with Descartes' argument. We might not yet have computer software to ensure that a system produces human-like behaviour displaying genuine originality or creativity, and Descartes might still maintain that such software is in principle impossible. But that would require an argument of a sort that he could not have contemplated. So far, then, he has not given us a compelling reason to suppose that bodily mechanisms alone are not enough for mentality.

However, he has another argument – philosophically more interesting – for holding not only that we cannot be machines, but also that minds are distinct from bodies and can exist without them. This depends on the assumption that if we can conceive of something 'clearly and distinctly', then that thing could exist exactly as we conceive of it, since 'all the things I conceive clearly and distinctly can be produced by God precisely as I conceive them' (Descartes 1641, *Meditation* 6). So, if I can clearly and distinctly conceive of a pig flying, then there could be flying pigs. This assumption is used to show that if I can clearly and distinctly conceive of my mind existing without my body, then my mind is 'distinct or different' from my body:

> it is sufficient for me to be able to conceive clearly and distinctly one thing without another, to be certain that the one is distinct or different from the other, because they can be placed in existence separately, at least by the omnipotence of God. (*Meditation* 6)

Another vital premiss is the conclusion of his famous 'cogito' argument. Because he thinks he can doubt everything except *that* he doubts, he cannot doubt he exists; indeed, he concludes, he knows for certain that he exists, and that nothing else belongs to his nature or essence except that he is 'a thinking thing' (given that doubting is a kind of thinking). He goes on to conclude that 'I, that is to say my mind, by which I am what I am, is entirely and truly distinct from my body, and may exist without it.' Summarizing:

(1) I clearly and distinctly conceive that I exist as a thinking thing.

(2) I can doubt that my body exists.

(3) So I can 'clearly and distinctly' conceive that I exist without my body.

(4) But if I can clearly and distinctly conceive 'one thing without another', then 'I am certain that the one is distinct or different from the other.'

(5) Therefore I am distinct from my body and can exist without it.

Not surprisingly, this argument has been subjected to heavy criticism (one objection focuses on the question of what is required for something to be 'conceivable'). But the key premiss is (4), and that is what I will focus on. A quick way to see there is something wrong with it is to consider that many people and things have more than one name and more than one description. The Roman politician Cicero was also an author, and had another name, 'Tully'. It seems that unless we already know that Cicero and Tully are one and the same person, we can clearly and distinctly conceive of the politician Cicero existing without the author Tully. But obviously it doesn't follow that Cicero is 'distinct or different' from Tully. Another much used example is of the Morning Star Phosphorus and the Evening Star Hesperus. It seems easy to conceive clearly and distinctly that Phosphorus should have existed without Hesperus also existing. Indeed, it took empirical work to establish that they are one and the same planet, Venus.

Those examples show that the vital premiss (4) of Descartes' argument is false, in which case the argument fails. Of course, that doesn't refute the view he is defending: Cartesian dualism. There may be other reasons to accept it, although I know of no good ones.

2. Against Cartesian dualism

Descartes' own contemporaries were greatly exercised by the question of how minds and bodies could interact. Minds are not extended in space, bodies are. Minds are supposed to be indivisible and to have no parts, bodies are divisible, with parts. How could such radically different things affect each other? This question prompted some odd theories. One was 'occasionalism', according to which neither minds nor bodies really have effects on anything, only God does. I may think that wanting a drink of coffee causes me to get up and make one, but really my wanting is just the 'occasion' for God to cause me to go to the kitchen. Another theory which dispensed with genuine mind–body interaction was 'pre-established harmony'. If one clock strikes the hour when another indicates precisely that hour, the explanation is not that the one *causes* the other to do what it does. Similarly, the fact that my wanting coffee is followed by preparing some is explained by the claim that my mental processes run in perfect pre-established coordination with my bodily activity. God has preset our minds and bodies to run in parallel, so there is always that kind of coordination.

Those theories are symptoms of desperation. Yet it is not clear that the worries prompting them ought to have troubled Descartes. If one is thinking about everyday cases of causation – matches caused to catch fire by being struck, water caused to boil by being heated – then it does seem that the cause is connected to the effect in a reasonably clear manner. But consider instead basic features of the world such as the behaviour of fundamental particles. It seems we just have to accept them: they are brute inexplicable facts. Comparably, Descartes could say it is just a basic inexplicable fact that minds and bodies interact.

A much more serious objection to his dualism – available to us but not to the science of his day – is that we now have compelling evidence that the brain is a purely physical system which – though deeply affected by its environment – determines behaviour and is not affected by anything non-physical. The neurosciences have built up increasingly detailed knowledge of the brain's workings, and there is overwhelming evidence that our cognitive and emotional activities depend on the brain and are liable to be affected in significant ways by whatever affects it, including drugs, diseases, and damage. Damage to certain areas damages memory; damage to other areas affects emotional responses – comparably for perception, planning, and the use of language. Certainly, much about the brain is still unknown, but a huge amount is already known about how its different components work: about the structure of brain cells and how neurotransmitters and neuroinhibitors facilitate or inhibit the transmission of electrochemical stimulation from neurone to neurone; about the different ways in which information is stored and retrieved; about perception, with a great deal of information about how sensory inputs are transmitted and have different effects in different regions of the brain; about sleep; about the effects of drugs; about the effects of disease – to mention only some examples. The further this research proceeds, the less plausible it is to suggest that anything non-physical affects what goes on in our brains, nor does there seem to be any shortage of neural resources for performing the functions involved in our mental lives. There are about 100 billion neurones and each sends and receives electrochemical signals from thousands of others – which provides a basis for storing, retrieving, and otherwise processing unimaginably vast quantities of information.

If Descartes were right, there would have to be gaps in the chains of physical cause and effect: places where the non-physical mind made its own special contribution to the causation of thought and action. Some physical events would have to be caused not by other physical events, but by non-physical mental events; there would also have to be physical events which did not cause other events in the body, but instead caused non-physical events in the mind. In fact, no such breaks in the causal processes of the brain have been discovered. Given their absence, together with the neuroscientific knowledge just alluded to, I take it that Descartes' theory has been refuted empirically. The decisive role of brains in our cognitive and emotional lives implies there is no work for Cartesian minds to do. For these and other reasons, Descartes' brand of dualism has few supporters today.

3. Epiphenomenalism and parallelism

But there are other varieties of dualism. The most significant of these accept that the brain both determines behaviour and is not affected by anything non-physical. Dualists of this sort accept the 'causal closure' of the physical: the view that every physical event has a physical cause. You might wonder why anyone willing to go that far would want to be a dualist at all, unless they just hanker for something even more mysterious than we know the physical world already is. But there is a more respectable motive, which many philosophers continue to find strong. They cannot see how *conscious experience* could possibly be explained in terms of a purely physical world; experiences seem utterly different in kind from anything the physical world contains. Since these dualists accept that all physical events are physically caused, they are forced to grit their teeth and say that the non-physical items in question, and with them consciousness itself, have no effects on the physical world. As T. H. Huxley (writing in the nineteenth century) put it, this view represents human beings as 'conscious automata'. According to it, the phenomena of consciousness are caused by purely physical processes, which also do the work of causing and explaining human behaviour. Conscious experiences themselves, on the other hand, are mere by-products of neural activity – they are causally inert 'epiphenomena' – and affect the physical world no more than the smoke from a factory chimney affects the processes which produce it. Those who hold this view – *epiphenomenalists* – usually hold that brain processes are sufficient for some kinds of mental activity, but not for consciousness. We shall take a closer look at epiphenomenalism later.

Parallelists also accept that every physical event has a physical cause and deny that consciousness has physical effects. But unlike epiphenomenalists, they deny that conscious experiences are caused by physical events. According to parallelists there is no causal relation at all between physical events and consciousness. Echoing the seventeenth-century views noted just now, parallelists require us to accept it as a brute fact that the events which constitute consciousness march in step with certain physical events – even though there is no causal relation between them.

You might think epiphenomenalism and parallelism are too odd to take seriously. Don't we know that experiences have physical effects? The alluring scent of freshly roasted coffee makes me want a cup, and I go ahead and boil some water. Didn't the experience of smelling the coffee *cause* me to go into the kitchen? But these modern dualists are painfully aware that conscious experiences *seem* to cause physical events. The trouble is that after serious and careful consideration they can see no way to make such causation intelligible. They conclude they must bite the bullet and maintain their counter-intuitive views in spite of their seeming absurdity.

4. Idealism

The core idea of idealism is that what exists depends on the mind. Recall the 'malicious demon' who, even if there had been no external world, might have arranged for all Descartes' experiences to have been just as they actually were (2§7). This unsettling thought was part of the reasoning which led Descartes to claim that he himself – at any rate his mind – was distinct from his body and could exist without it. All of which suggests an unsettling question: why do we have to suppose there really is an external world? His answer, roughly, was that God would never allow us to be so deceived in such an important matter. However, some people went on to suggest that Descartes and most of the rest of us are mistaken about the nature of reality. There really is no external world of the sort we assume, and it is God who has arranged for our experiences to be as they are – in the absence of such an external world.

George Berkeley (1685–1753) was a powerful and influential exponent of such idealism. He maintained that the existence of 'houses, mountains, rivers, and in a word all sensible objects' actually consists in their being seen, felt, or otherwise perceived, and that the contrary view that they exist independently of being perceived involves 'a manifest contradiction' (1710: I.4). He based his position on the assumption that all we ever perceive is ideas or sensations:

> The table I write on, I say, exists; that is, I see and feel it: and if I were out of my study I should say it existed; meaning thereby that if I was in my study I might perceive it, or that some other spirit actually does perceive it.' (1710: I.3)

So according to Berkeley there is no more to the existence of his table than either his actually seeing or feeling it, or else the fact that if he were to be in a suitable position he would see or feel it: in his vocabulary, he would have appropriate 'ideas'. So for him, all that exists is God (an infinite spirit) ourselves (finite spirits) and ideas (conscious experiences).

Since the main focus of this book is the nature of consciousness and the kind of functionalist approach that I think helps us to understand it, we need not go more deeply into idealism; but it is an important part of the background.

5. Dual aspect theory

Spinoza (1632–77) argued that God and Nature are not different entities, but one and the same. Analogously he argued that mind and body are not, as Descartes held, two distinct substances, but a single individual with two 'aspects'. Under one aspect this individual is an 'idea', under the other it is a body. If sense can be made of these notions, the dual aspect theory offers a satisfying solution to the problem of how physical and mental items can have effects on each other. If mind and body are just two aspects of the same thing, then whatever either of them causes is also caused by the other: their interaction raises no difficulties.

But can we really make sense of the dual aspect theory? It would have been clear enough if the idea had been simply that the mental vocabulary is just a special way of talking about things specifiable by means of the physical vocabulary, and vice versa. I am happy with the first half of that claim: I accept that the mental vocabulary is indeed a way of talking about (parts of) a purely physical world. In fact, that is a partial statement of physicalism (only partial, because it needs the addition of a clause to ensure that what the physical vocabulary specifies is fundamental). The other half of the claim, that the physical vocabulary is a way of talking about the mental world, also seems clear, at least on the face of it. But as soon as we try to discover how it is supposed to apply to reality, we find that this component of the dual aspect doctrine is extremely hard to understand. It implies that descriptions of the sizes and masses of stars and planets, for example, are just ways of talking about *mental* states. But does that make sense? Evidently a lot of special explanations are needed before we can even understand what the dual aspect theory actually is: attractive when you first come across it; increasingly mysterious on closer acquaintance.

6. Panpsychism

An ancient idea in some ways similar to the dual aspect theory is that consciousness somehow pervades the whole of nature. Two main versions of this approach are taken seriously by some philosophers today. One, which I will call 'standard panpsychism', is that there is a sort of non-physical 'mind dust' everywhere – even in the elementary particles of physics. The other is 'Russellian monism' (a version of it having been suggested by Bertrand Russell).

Standard panpsychists, like dualists and many others, are impressed by the difference between actually having conscious experiences – for example, having toothache or smelling eucalyptus – and physical facts about the workings of bodies and brains: consciousness and the physical just seem totally different kinds of things. They reason as follows: either consciousness has been around ever since the beginning of the universe or else it has somehow come into existence – presumably when appropriately equipped organisms had evolved. But if the latter, there is a huge problem: how could something so special and apparently so different have come into existence from the merely physical? Finding no plausible answer, panpsychists conclude it has always been around – and emphasize that in their opinion it would have been impossible for the mental to have originated from the non-mental. Since science tells us that everything is made up of the same fundamental components (leptons, baryons, strings, or whatever it might be), this seems to imply that all these universally pervasive fundamental items have consciousness in them, or perhaps only the 'seeds' of consciousness.

Does that mean things such as sticks and stones are conscious? Panpsychists typically reject that interpretation. One relatively sober version of their view is that although not every physical object is conscious, everything is *composed* of items which constitute something conscious in appropriate combinations. There are big problems with this view, however. To mention one: how could the mere assembly of a number of centres of potential consciousness constitute a single conscious being? Does a crowd of people, each of whom is individually conscious, constitute another consciousness, a single overarching one? How could it? That is a very hard question for panpsychists. And what would determine the nature of the experiences such a collective centre of consciousness would have? Another hard question.

Let us turn to Russellian monism, whose special features are worth noting. Its exponents typically start by accepting that normal physics provides a true picture of parts of reality. They typically claim that there is only one fundamental class of things in the world – they are indeed monists – and they often say these things are 'physical'. But they go on to say that normal physics leaves out something absolutely crucial: it tells us only about the structures of physical objects and their causal and other relations to one another, not about their intrinsic or categorical nature. Physical theory gives us equations specifying the functions performed by properties such as mass, charge, spin, and so on, together with how they are related to each other. But what about the categorical nature of these items? Suppose we were interested in a couple called Anna and Bill, and our informants gave us nothing but descriptions of their ages, heights, weights, where they lived, how they interacted, what other people they knew, where they worked, and other relational facts. That wouldn't satisfy us. We'd still want to know what they were actually like, what sort of people they were. Comparably, according to these panpsychists, physics leaves us in the dark about the inner natures of the entities figuring in its descriptions and explanations of things in the world.

So, what are these categorical properties? According to Russellian monists they are 'phenomenal': that is they provide consciousness. (I shall use this word 'phenomenal' quite often from now on; it is a convenient way to signal that we are thinking about the actual character of conscious experiences: *what they are like*.)

By emphasizing what they conceive of as the twofold character of natural objects – their purely structural and relational properties on the one hand and their intrinsic and supposedly phenomenal properties on the other – panpsychists commit themselves to a kind of dualism, and this lays their position open to standard objections. In particular, they face the following dilemma. Do they maintain that the phenomenal properties of objects have effects on their structural properties, and in that way affect the ordinary physical world? Then their view is in conflict with the well-supported thesis that every physical effect has a physical cause – and not a non-physical one. If however they choose the other alternative and say the allegedly non-physical phenomenal properties have no physical effects, then they incur the difficulties of epiphenomenalism, which will be discussed in more detail later.

Here is another worrying question for panpsychists. Do they think it is possible that the world should have been as it actually is in those respects which are described and explained by physics, but altogether without phenomenal properties, so that the organisms which corresponded to us had no conscious experiences at all? If so, such a world would be a zombie world, and these panpsychists would be vulnerable to arguments against the possibility of such worlds (one such argument will be set out in Chapter 7). You might suggest they could avoid that difficulty by accepting that there could not be zombie worlds: by maintaining it is simply impossible that the world should have been physically as it actually is while its inhabitants were not conscious. But in that case the physical facts alone would have been sufficient for consciousness – which would make these thinkers physicalists, inconsistently with their view that consciousness depends on something utterly different from the physical.

It seems to me that panpsychist reasoning only looks plausible if you start by presupposing that consciousness is a thing or a sort of stuff: like a jacket or paint, which an object might have or be without. In the second half of this book I hope it will become clear that nothing like this can possibly be the case.

7. Physicalism

According to the ancient Greek version of the atomic theory, all that exists is atoms and empty space. Atoms ('uncuttables') were said to be solid particles whose different sizes and shapes (rounded, hooked, pointed, and so on) resulted in their combining in numerous different ways to make earth, air, fire, water, and the other substances we find

IS SOMETHING NON-PHYSICAL INVOLVED?

in the world. Fire atoms, for example, had sharp points, so they could penetrate and break up things like pieces of wood. That was supposed to explain combustion, and how wood (which was supposed to consist of different kinds of atoms in a particular configuration) comes to be replaced, as it burns, by smoke and ashes, which were different arrangements of some of the original atoms composing the wood. A human soul, in contrast, consisted of many tiny smooth round atoms, small enough to pass through the gaps between the larger atoms composing the body. Their motions caused our limbs to make the movements constituting our behaviour.

That was an early version of *materialism*. When originally proposed, it was only a minority theory, but it has survived in some form or other for two and a half milennia, Thomas Hobbes in the seventeenth century being a more recent exponent. Today almost all scientists and most philosophers seem to be materialists of some sort. However, since modern physics tells us that the fundamental components of things are nothing like the solid particles described by the ancient atomists and Hobbes, the corresponding view is often labelled 'physicalism'. The idea is that nothing exists but the physical, which is taken to be, broadly, whatever physics tells us about.

In spite of its wide acceptance by scientists and philosophers, physicalism faces big problems in connection with consciousness. How could organisms composed of atoms, electrons, and other particles, no matter how complicated they might be, produce what on the face of it seems to be a wholly different category of things: conscious experiences? For a start, it clearly doesn't make sense to suppose that experiences are *made from* atoms and the rest. And since there seems to be a huge difference between the churnings of the neurones on the one hand, and the flow of experiences on the other, it is not surprising that many people still find dualism appealing.

Yet dualism only looks like an answer if you have given up. It goes no way towards *explaining* consciousness. It simply delivers the blank assertion that the phenomena of consciousness are a fundamental component of what exists: brute facts to be accepted but not explained. Dualists and idealists tend to criticize physicalists for failing to come up with plausible explanations of how there can be conscious experiences if we are purely physical beings – yet they themselves have no explanations at all, plausible or not.

8. The identity theory

The most popular version of physicalism over many years has been the *identity theory* (or the 'psycho-physical identity theory'), according to which experiences are simply the same things as certain physical processes – brain processes, no doubt. One great advantage this theory has over dualism is that it raises no difficulties about causation. If experiences are identical with brain processes, then since the latter are unproblematically both causes and effects of other physical processes, so are experiences. But the identity theory has its own problems.

One is that it is implausible to say each type of experience is identical with just one type of brain process. We have no difficulty ascribing pains, for example, to a whole range of different creatures, not just human beings – when these other creatures have different kinds of nervous systems. Similarly for individual human beings. Brain damage can result in the formation of new neural pathways, so that the same type of experience in the same person is underlain by different brain processes. Given those facts, pain cannot be identical with one particular type of brain process. This is the so-called 'multiple realizability' objection to the identity theory.

Many identity theorists retreat from the claim that each *type* of experience is identical with some type of brain process, and say only that each particular *instance* of an experience is identical with some physical process. Many people assume that this is what physicalism is. However, this view too is open to objections.

One is that although physicalists claim there is nothing beyond the purely physical, the bald statement that something x is identical with something y does not give special priority to either x or y. If I say the Morning Star is identical with the Evening Star, that doesn't imply the Morning Star is somehow more fundamental or basic than the Evening Star: it leaves them on a level – when the physicalistic identity theory requires the physical to be fundamental. J. J. C. Smart was a highly influential physicalist who famously claimed that *sensations are brain processes* (Smart 1959). However, that statement does not by itself express his physicalism for the reason just noted: it does not imply that sensations depend on brain processes rather than that the latter depend on sensations. However, Smart didn't think the identity theory amounted to a commitment to physicalism all by itself. His point was that if physicalists are pressed to say what sensations and other experiences are, a good reply is that they are brain processes: there is no more to them than that. The identity thesis by itself is no good as a statement of physicalism; it needs to be backed up by a clause on the lines of:

(1) Everything is purely physical

The snag now (as you will have noticed) is that (1) is itself a version of physicalism. Since the identity thesis doesn't commit you to (1) or anything like it, it doesn't commit you to physicalism. But if we attempt to fill the gap by adding something like (1), the identity thesis itself becomes redundant.

Another objection raised against the identity thesis as a statement of physicalism is that *being a certain brain process* is a different sort of thing – a different property – from (for example) *being the experience you have when you see something blue*. We can know perfectly well what it is like to have the blue experience without having any inkling of any underlying brain processes; similarly, we may be able to discover which brain processes are going on in someone's head without knowing what experience, if any, those processes underlie (recall Nagel's bats). A partial reply to this objection is that what seem to be two different properties can turn out to be one and the same property under different descriptions. For example, the property of having two ears is the same as the property of having the number of one's ears equal to the cube root of 8. But even if we concede for argument's sake that being such-and-such a brain process and being the experience you have when you see something blue are in fact one and the same property, a serious problem remains. What determines that one sort of brain process rather than another is indeed identical with an experience in the first place? Plenty of brain processes do *not* underlie conscious experiences; so what makes the difference between those that do and those that don't? On that question the identity theory is silent.

Over the years, one path out of that difficulty – along a route broadly indicated by Smart himself – has gained more and more followers: functionalism, about which I shall be saying a lot later. As we shall see, if functionalism is true, then (assuming the physical universe is as physicalists suppose) there is no need to establish psycho-physical identities. It is enough to show that the relevant functions are in fact performed by purely physical items. And functionalism is clearly not exposed to the multiple realizability objection: indeed, it is the thesis that mental states actually are multiply realizable.

Still, many philosophers think that the difficulties surrounding physicalism cannot be overcome. They don't see how the resources of a purely physical world could be sufficient to explain consciousness, and conclude that something non-physical must be involved. That brings us to what many regard as a big objection to all varieties of physicalism, not just the identity theory: the idea of (philosophical) zombies, to be examined in the next two chapters.

9. Conclusion

It is a striking fact that dualism, idealism, dual aspect theory, panpsychism, and physicalism offer no explanation of consciousness. They say only that consciousness is non-physical, or an aspect of something special, or distributed over all objects, or physical. But – odd though this may seem – these theories are beside the point, given that the point (for us at any rate) is to try to understand *what it takes* for something to be conscious. Physicalists as much as dualists and the rest still need to provide explanations and understanding. And while dualism, idealism, and panpsychism say only that mentality is a basic, brute, and inexplicable feature of the universe, physicalism – although it does not by itself provide an explanation – leaves scope for one; it doesn't just shut off the possibility.

6

Zombies

1. Introduction

Our guiding question is: what does it take for something to be conscious? Many people suggest that the only way to answer this question is through science: psychology, neuropsychology, neurobiology, and the rest. I personally am persuaded that science is our best source of basic knowledge about the world – with emphasis on 'basic': of course, we have other sources of knowledge as well, including personal experience. The trouble is that although we must take full account of scientific results – suggestions for explaining consciousness had better not actually conflict with the science – it has become clear that scientific work alone won't be enough; philosophical work is necessary as well. A legal case illustrates the point. A cook put live prawns on a hot plate, where they squirmed and wriggled, apparently in pain. The cook was charged with cruelty to animals, but the case was dropped because expert advice on whether prawns could feel pain could not be found. The prawns' behaviour made it easy to think they were suffering, but perhaps nothing was involved over and above their behaviour: maybe they had no sensations at all – any more than a twisted rubber band, writhing as it unwinds, has sensations.

You might think that further research into these creatures' nervous systems would settle the matter. There is indeed research aiming to discover the neural processes *correlated* with consciousness. One suggestion has been that it is correlated with the synchronized firing, at a frequency of 40 hertz, of neurones connecting the thalamus and the cortex (this idea was defended by Francis Crick, the Nobel prizewinning co-discoverer of the structure of DNA). But such suggestions leave the fundamental question unanswered. They fail to explain what it is about that particular kind of neural activity which ensures it is both necessary and sufficient for consciousness. Suggested answers to this question appeal to the functions performed by the activity in question, for example that it is involved in attention. But although the answers may be correct, they fail to say whether performance of the suggested functions is sufficient for consciousness. Suppose, then, that the relevant purely scientific work has been done and on that basis it is claimed that prawns can feel pain. Wouldn't it still be possible to deny this and maintain they lacked all sensation, and that what appeared to be pain behaviour was not after all an effect of pain, but of non-conscious processes inside their nervous systems?

2. The zombie idea

If that suggestion seems absurd, recall the striking theory mentioned earlier: epiphenomenalism. Scientific developments in the nineteenth century, for example the discovery of electromagnetism and the beginnings of neurophysiology, encouraged the idea that physics was capable of explaining every physical occurrence that was explicable at all. It seemed that every physical event had a physical cause: that the physical world was 'closed under causation'. Some people expected that if neurophysiology came up to expectation, then even human behaviour could be explained in those terms. But how was consciousness supposed to fit into the story? Thoroughgoing materialists insisted that nothing was going on but physical processes (hence the psycho-physical identity theory). Yet experiences certainly appear to be different in kind from anything physical, and others concluded that consciousness must involve something non-physical. But these thinkers then faced another difficulty, as we have seen. Still assuming that all physical events are explicable physically, they had to conclude that consciousness has no effects on the physical world, in which case human beings are 'conscious automata'. Our behaviour must result from the purely physical processes going on in our bodies as they interact with the outside world, and consciousness can only be an inert non-physical by-product. When you follow that idea through, you realize it means that – just possibly – there could have been purely physical duplicates of ourselves, organisms just like us (as physicalists conceive of us) in every physical detail, yet without consciousness.

As G. F. Stout said, if epiphenomenalism is right, then:

> it ought to be quite credible that the constitution and course of nature would be otherwise just the same as it is if there were not and never had been any experiencing individuals. Human bodies would still have gone through the motions of making and using bridges, telephones and telegraphs, of writing and reading books, of speaking in Parliament, of arguing about materialism, and so on. (1931: 138f)

What Stout describes here is a world of (philosophical) zombies. When I started teaching philosophy I took a modified behaviouristic or Rylean line. A first year student put an end to those dogmatic slumbers. 'There are zombies, aren't there?' she asked. I thought that made her look pretty simple, but really I was the simpleton. I had no plausible reply, and it was only later that I came to realize she had a point. She was probably thinking of the creatures represented in zombie films, or of the zombies of West Indian folklore – corpses magically caused to work for wizards. I wasn't inclined to believe there were such things. In any case, even if there were zombies of that folkloric kind, their behaviour would be caused by others, and we should have no reason to suppose they themselves would be conscious. We don't think marionettes or glove puppets or even the Giant are conscious for the excellent reason that we know that their apparently intelligent and sentient behaviour is caused by others. The zombies that cause trouble in philosophy are different. They are exact physical duplicates of ourselves, inside and out, yet still not conscious. Whatever you may think about that idea, it forces you to think deeply about consciousness. It also helps to make clear that scientific research by itself will not be enough to answer the question of what it takes for something to be conscious.

A key consideration is that the zombie idea requires us to accept at least the possibility that consciousness has no effects on the physical world. For if zombies are possible, then what we tend to think of as effects of our experiences – for example, saying things like 'Did you see how green that meteor was?', or 'Eucalyptus always makes me think of koalas' – are nothing of the sort; instead they are the effects of purely physical processes churning on independently: the experiences themselves make no contribution. Many people reject that consequence of the zombie idea. As Stout put it, 'There can be no doubt that this is *prima facie* incredible to Common Sense' (1931: 138). But just finding the idea of zombies hard to swallow is not enough – and far from proving it is altogether mistaken.

Epiphenomenalists are well aware that their theory is inconsistent with common sense assumptions. But they point out that plenty of things that are prima facie incredible have been discovered to be actually the case. Think how much work it took to bring people round to the idea that the earth moves round the sun, rather than vice versa. It does indeed seem incredible that our experiences should have nothing to do with our behaviour. But epiphenomenalists think they have overwhelming reasons for their theory. On the one hand they accept that all physical events have purely physical causes. On the other they cannot see how something so apparently distinct from the physical world as consciousness should be, at bottom, purely physical. They are powerfully influenced by the idea of zombies, and more generally by the idea that the narrowly physical facts seem to leave room for the facts of consciousness to have been different from what they actually are. Clearly, if zombies are so much as possible, then the mental facts about our world – which in that case depend on something non-physical – cannot be explained in purely physical terms. So it is not enough to point to the seeming absurdity of the consequences of epiphenomenalism. It is necessary to prove that an epiphenomenalistic world is absolutely impossible – which is just what I shall try to do in the next chapter. But we need to be clear about what kind of possibility matters.

3. Logical possibility and impossibility

There is natural possibility and there is logical possibility. (There are other kinds too, and a great deal of literature on the subject, but what follows should be enough for our purposes.) A state of affairs is *naturally* possible if it is not ruled out by the laws of nature. Given those laws, it is naturally possible to dissolve ordinary sugar in water, and naturally impossible for a solid ball of lead to float on water. But something may be *logically* possible even if it conflicts with the laws of nature. It is logically possible for a solid ball of lead to float on water, and for sugar not to dissolve. In the arguments we shall be discussing, it is logical possibility and impossibility that will concern us.

The most obvious cases of logical impossibility involve explicit contradictions, as in 'It's raining here now and it isn't raining here now.' We see immediately that they could not possibly be true – and the laws of nature have nothing to do with it. But some contradictions are not explicit, and hard to discover. 'Square equal in area to a given circle and constructed with ruler and compasses alone' is a description which doesn't involve any obvious contradiction, and many people (among them Thomas Hobbes) tried hard to find a construction satisfying it; in that sense they tried to square the circle. But eventually it was proved that such a construction is logically impossible.

Now, together with most people I assume that the laws of nature do not permit the existence of zombies: such creatures are at any rate naturally impossible. But that doesn't raise any special difficulty. When philosophical zombies are said to be possible, the idea is not that such creatures might really be roaming the earth. Even dualists, who believe the world includes non-physical items as well as physical ones, tend to agree that nature ensures that the physical facts bring with them the special non-physical items they think are necessary for consciousness. The claim made by those who think zombies are possible is simply that *the idea of zombies doesn't involve a contradiction:* zombies are logically possible in that broad sense. If they are indeed logically possible in that sense, then clearly, because they are supposed to have all our own purely physical properties without consciousness, the purely physical facts about us cannot be enough to explain consciousness. If it is logically possible that there should have been an exact duplicate of the physical world where our counterparts lacked consciousness, then something beyond the merely physical is needed to explain it.

But if something non-physical is involved in consciousness, it seems we have to regard it as a basic feature of our universe in addition to the swirl of matter and energy that science tells us about. All we can be supposed to know about this 'something' is that it provides for conscious experiences, so that when it is added to the purely physical set-up in a zombie world, it ensures that what would otherwise have been mere zombies are conscious. Now, we know the purely physical world is full of amazing things, such as black holes and dark matter. Why shouldn't the facts of consciousness involve similarly extraordinary and counter-intuitive things? It is not good enough to dismiss the zombie idea by pointing out that it conflicts with ordinary assumptions; arguments are needed. But first let us consider what can be said for the idea that zombies are logically possible; I will not argue against it till the next chapter.

4. Arguments for the logical possibility of zombies

On first thinking about that idea, and (believe me) on thinking about it carefully for a long time, it doesn't seem to involve any contradiction. But when I first put it to my colleagues they rightly objected that I couldn't just assume there is no hidden contradiction, I must prove it. That was annoying – it seemed just obvious that zombies were logically possible – but eventually I came up with a couple of arguments which seemed at that time to do what was necessary (Kirk 1974a,b). A glance at them may be helpful.

Dan. One of the arguments describes Dan, whose behaviour starts to show features which suggest that at six-monthly intervals he is losing each of his senses one by one – but in a bizarre way. The first symptoms were that, in spite of being able to do everything he could do before, including driving his car and reading books, he protested that he'd suddenly become blind and wasn't really able to see anything. After six months he displayed similar behaviour in connection with hearing: said he had gone deaf, yet still took part in conversations and listened to music with apparent enjoyment. The same happened with each of his other senses, one by one, although he seemed to find it increasingly difficult to register his protests. When eventually all his senses had been affected, his patterns of behaviour reverted to normal and the protests ceased. However, some of his friends suggested that at that point he had in reality *lost* all conscious sensory experience in spite of behaving like a normal person: that he had become a zombie.

The claim is not that this story rules out alternative explanations, but that it makes it intelligible to say Dan had become a zombie, and shows that no contradiction is involved in the idea. Is that reasoning sound?

No. The story does not prove there is no contradiction in the idea of a zombie, but at most only that we can be persuaded there is no contradiction. The story is just an intuition pump – and misleading. Consider what must be supposed to happen in Dan's body. The picture we are given is that while physical processes alone continue to cause behaviour consistent with normal conscious experiences, he himself knows that – in each of the sensory domains so far affected – he is *not* having any such experiences. So in the course of the successive changes described in the story, we have to suppose his behaviour has two distinct sets of causes. One is the purely physical processes which supposedly explain his apparently normal behaviour. The other is whatever it is which causes him to protest that he is really blind, deaf, and the rest. But this second source of behaviour cannot be physical, since the story requires it to interfere with the normal course of physical causation. In fact this second source of behaviour has to be supposed to be its normal cause prior to the onset of the changes. But since at the same time purely physical processes are supposed to continue to cause the same normal behaviour ('It smells nice', 'Look at the sunset'), the story depends on inconsistent assumptions. On the one hand Dan's normal behaviour has to be caused purely physically; on the other, it has to be caused non-physically. Far from proving that the idea of zombies involves no contradiction, therefore, the story itself is internally inconsistent.

Zulliver. Another thought experiment builds on the story of Gulliver in Lilliput. The creatures he comes across are tinier than the ones Swift imagined. These micro-Lilliputians get inside Gulliver's head and interfere with his brain. First they disconnect the afferent nerves from his sense organs and the efferent ones to his muscles. Then they themselves monitor what is transmitted from the former and send outputs down the latter, causing behaviour indistinguishable from what it would have been originally. The result is Zulliver, a system with the same behavioural dispositions as Gulliver but no conscious experiences – or so the argument runs. Consider, though, what the argument is intended to show: that the purely physical facts about a system cannot logically guarantee that it is conscious. In order for this argument to work, it has to be assumed that Zulliver is *not* conscious – indeed that is taken to be obvious – and the argument is a challenge to those who maintain that the purely physical facts about Gulliver logically guarantee his consciousness. After all, the physical processes inside his head don't do anything significantly different from the processes inside Zulliver's head. If the ones inside Zulliver's head aren't enough to make him conscious, what is it about those inside Gulliver's head which are, logically, enough to make *him* conscious? On the face of it, there isn't any rational principle that would enable us to settle the matter.

At any rate that was the argument. I have already mentioned one defect: it simply presupposes (not unreasonably) that Zulliver is not conscious. Later in the book I will argue that there are rational principles which enable us to settle the matter: to see that the argument fails decisively.

5. Chalmers's first argument

Over a number of years, David Chalmers has argued powerfully that zombies are logically possible; it will be useful to consider his five main arguments. The first takes the micro-Lilliputian idea further. There are huge numbers of homunculi, each of whom performs the functions of an individual neurone: some receive signals from the afferent nerve endings in your head and transmit them to colleagues by mobile phone; others receive signals from colleagues and transmit them to your efferent nerves; most receive and send signals to colleagues in ways that mirror the activities of our neurones. Would such a system be conscious? Chalmers claims it might not be. Note to start with that the argument doesn't require us to go so far as to conclude that the system would definitely *not* be conscious. It depends only on the assumption that its not being conscious is *conceivable*, which many find reasonable. In Chalmers's words, all that matters here is that when we say the system might lack consciousness, 'a meaningful possibility is being expressed, and it is an open question whether consciousness arises or not' (1996: 97). Possibly, then – as we might suppose for argument's sake – the system is not conscious. But if it isn't, then it is already very much like a zombie. The only difference is that it has little people where a zombie has neurones. And why should that difference be relevant? Why should it be logically necessary that switching from homunculi to neurones switches on the light of consciousness?

Like the argument from Zulliver, this leaves us wondering whether there are any rational principles we can appeal to in order to answer the challenge. As I have said, what I believe to be such rational principles will be formulated later.

6. Chalmers's second argument

Chalmers goes on to appeal to the old idea that you and I might experience colours in systematically different ways, where the differences were systematically coordinated so as to make them undetectable. A suggestion of John Locke's makes a good starting point:

> Neither would it carry any Imputation of *Falshood* to our simple *Ideas*, if by the different structure of our Organs, it were so ordered, That *the same Object should produce in several Men's Minds different* Ideas at the same time; *v. g.* if the *Idea*, that a *Violet* produced in one Man's Mind by his Eyes, were the same that a *Marigold* produced in another Man's, and *vice versa*. For since this could never be known ...; neither the Ideas hereby, nor the Names, would be at all confounded, or any *Falshood* be in either. (*Essay*, II, xxxii, 15)

The suggestion is that the experience you have when we are both seeing the same blue object in the same situation might be the experience I should have had if I had been seeing an orange object in that same situation; and vice versa; and so on systematically through the spectrum, respecting complementarity. Assuming for the sake of argument that the colour solid is symmetrical (although it isn't), and that these differences between us had existed from birth, they would not show up in our behaviour. In particular, we should agree in our use of words, since each would have learnt to call blue things 'blue', orange things 'orange', and so on, in spite of the differences between our subjective experiences of colours. This is the idea of the 'inverted spectrum' or 'transposed qualia'.

Now, the only way this argument will do the work Chalmers needs is if the differences in question could exist even if there were no *physical* differences between us. If the imagined situation depended on physical differences, then it would not show that consciousness depended on something non-physical and would therefore not support the case for the zombie possibility. But if he just presupposes that the example shows something non-physical is necessary for consciousness, he is begging the question. He says, 'it seems entirely coherent that experiences could be inverted while physical structure is duplicated exactly' (1996: 100). Yes, it *seems* coherent; but we already knew that. We also knew that such seeming doesn't amount to an argument. So this thought experiment is another appeal to untutored assumptions: a mere intuition pump.

7. Chalmers's third argument

Even if we knew every last detail about the physics of the universe – the configuration, causation, and evolution among all the fields and particles in the spatiotemporal manifold – *that* information would not lead us to postulate the existence of conscious experience. (Chalmers 1996: 101)

Chalmers is right to say the information he mentions would not necessarily lead anyone to ascribe conscious experiences to physical organisms. But is that relevant? What matters is whether the physical facts do or do not leave open the logical possibility that physical organisms such as ourselves might fail to be conscious – regardless of what anyone may be led to think. His aim is to establish that they do leave that possibility open, so his quoted remark gives no support to the claim: it just asserts it. He concedes that the physical facts about the world 'might provide some indirect evidence for the existence of consciousness'; some organisms themselves might even claim to be conscious. But he asserts that 'this evidence would be quite inconclusive, and it might be most natural to draw an eliminativist conclusion – that there was in fact no *experience* present in these creatures, just a lot of talk' (1996: 102). As before, though, while those claims may be intuitively appealing, they fall well short of proof.

8. Chalmers's fourth argument

He goes on to appeal to a famous argument from Frank Jackson: the 'knowledge argument' (1982). This is based on a thought experiment about Mary, a superb scientist who has been brought up in an environment free of all colours but black, white, and shades of grey. She knows the physical facts about colour vision but – never having actually seen anything coloured – does not know what the experience of seeing red is like; she discovers that only after she has been let out. Chalmers claims 'No amount of reasoning from the physical facts alone will give her this knowledge', and that 'it follows that the facts about the subjective experience of colour vision are not entailed by the physical facts' (1996: 103).

Does it really follow? We can accept that Mary doesn't know what it's like to see red before she has had actual experiences of red things. But think why that seems plausible. It is that in order to know what it is like to see a colour you must have had some experience of it, or a similar colour, or at any rate you must be able to construct an appropriate experience in imagination. Mary hasn't had such experiences, and that is why she is not in a position to infer, from the physical facts alone, what it's like to see red things. But that doesn't mean the truths in question 'are not entailed by the physical facts'. Here an important distinction is being overlooked. The question of what can be *known* on the basis of the physical facts is not the same as the question of what actually *follows from* (is entailed by) those facts. I shall be arguing that the facts about subjective experience follow from the physical facts. But at the same time I agree with Jackson that Mary's knowledge of the relevant physical facts is not enough to enable her to know what it is like to see red. Since Chalmers lumps those two things together, I don't think his present argument works. (These thoughts will be followed up in Chapter 12.)

9. Chalmers's fifth argument

The fifth and last argument Chalmers offers is 'from the absence of analysis'. He rightly points out that his opponents 'will have to give us some idea of how the existence of consciousness *might* be entailed by physical facts', but goes on to assert that 'any attempt to demonstrate such an entailment is doomed to failure' (1996: 104). If that is right, physicalism is sunk. But his reasons for his claim fall far short of justifying it. He starts from the plausible assertion that 'the only analysis of consciousness that seems even remotely tenable for these purposes is a functional analysis' (one example of which will be developed later in this book). He then asserts that what makes states conscious 'is that they have a certain phenomenal feel, and this feel is not something that can be functionally defined away' (1996: 105). (By 'phenomenal feel' he means what it is like actually to be in the conscious states in question.) The trouble is that his last statement is no more than an assertion of what his opponents, including me, deny. It doesn't amount to an argument. (See 10§4.)

10. Conclusion

The arguments for the logical possibility of zombies examined in this chapter are interesting, and some of them can seem strong. Though not conclusive, they pose a serious challenge to those who maintain that zombies are impossible. In any case, to show that certain arguments don't work is not to show that the claim they are supposed to support is false: the accused may be found not guilty even when they actually did the deed. On the basis of what has been said so far, we are not entitled to conclude that zombies are impossible, only that their possibility has not yet been proved. And the arguments from Zulliver and neurone-mimicking homunculi leave one wondering whether there are any rational principles that would settle the matter.

7

What's wrong with the zombie idea?

1. Introduction

A zombie world would be an exact physical duplicate of our world as physicalists conceive of it. It would contain nothing but physical things subject to physical laws, and the behaviour of its human-like inhabitants would be caused by purely physical processes – but they would have no conscious experiences. To use jargon, they would have no 'qualia' (Latin plural: singular 'quale') – a word I shall be using as shorthand for 'the non-physical components of conscious experience as conceived of by epiphenomenalism'.

The question is whether that is possible. At this stage you may not have a definite opinion, but it is certain that any serious philosophical investigation of consciousness must go into the matter. The possibility in question is what I am calling logical, rather than mere imaginability or mere consistency with the laws of nature: the description of a zombie world must involve no contradiction. And the vital point now is that if a zombie world really is possible in that sense, then consciousness cannot be explained on the basis of the physical facts alone, and we are forced into some variety of dualism. As we saw earlier, dualism requires us to accept that consciousness is a brute fact with no explanation. If on the other hand zombies are *not* logically possible, then we can forget about the dispute between dualism and physicalism and concentrate on the question that remains just as deeply puzzling as ever: how on earth can there be such a thing as consciousness at all?

I have become more and more firmly convinced over the years that the zombie idea, in spite of its allure, springs from fundamentally mistaken assumptions – though these are not obvious. I shall call those who hold that zombies are logically possible 'zombists', and argue that if zombists had explored the implications of their view deeply enough, they would have seen that it involves a contradiction. My aim in this chapter is to make clear to any reader who has no axe to grind, and is willing to follow some fairly detailed reasoning, that zombies are not logically possible. This reasoning will put us on the way to explaining what it takes for something to be conscious.

2. Epiphenomenalism again

The main argument has two parts. The first shows that epiphenomenalism is not logically possible. The second shows that if zombies had been logically possible, then epiphenomenalism would have been possible too. Given the first part, it follows that zombies are not logically possible.

Epiphenomenalism, you may recall, is the view that our world might have been such that each of us had two components: a body whose purely physical workings accounted for all our behaviour, and a non-physical component, qualia, which accounted for consciousness. The non-physical component would be caused by physical events but – here is the snag – it would have no effects on the physical world. I shall not be arguing that zombists must accept that epiphenomenalism is *actually* true of our world; only that they must accept it is possibly true.

In more detail, epiphenomenalism commits you to the following five theses.

(1) The world is partly physical and its physical component is as conceived of by physicalists, so that every physical effect has a physical cause.

(2) Human beings are physical systems related to a special kind of non-physical items (non-physical 'qualia') and it is the latter which provide for conscious experiences.

(3) These non-physical qualia are wholly caused by physical processes but have no physical effects.

(4) Human beings consist of nothing but functioning bodies and non-physical qualia.

(5) Human beings can notice, attend to, think about, compare, and (on occasions) remember features of their non-physical qualia.

I describe these epiphenomenalistic qualia as non-physical because many physicalists, including me, agree we have qualia in a broader sense – because there is *something it is like* for us to have experiences – even though we are purely physical organisms.

I shall argue that all zombists, regardless of whether they are also epiphenomenalists, are committed to the view that a world satisfying (1)–(5) is logically possible. If that is right, it doesn't matter if there are alternative versions of epiphenomenalism; if a world satisfying this particular set of conditions, (1)–(5), is not possible, then it will follow that a zombie world is not possible either. In outline:

(a) If a zombie world is possible, so is a world satisfying (1)–(5).

(b) A world satisfying (1)–(5) is not possible.

(c) Therefore, a zombie world is not possible.

The trouble with epiphenomenalism is that (5) is inconsistent with (1)–(4). As I hope to make clear, we could notice, attend to, think about, compare, and remember things about qualia only if qualia had physical effects – when it follows from (1)–(4) that they have no physical effects. Thus, the truth of (1)–(4) would rule out the truth of (5). (For further details of these arguments see Kirk [2005] 2008 and 2008.)

3. Epiphenomenalism's big problem

We know what our experiences are like – hearing the cuckoo, for example, or feeling a twinge of toothache. We can attend to them and, on occasions, remember them and compare them with one another.

No one denies those facts, least of all epiphenomenalists, since otherwise they couldn't state their theory. But does epiphenomenalism allow those facts to be explained? There would not have been much difficulty if it were agreed that experiences *cause* behaviour and other physical events, but that is just what epiphenomenalism rules out. If shouting 'Ow!' at the dentist's were actually caused by pain, then that, together with similar facts (such as the taste of your coffee causing me to say how nice it is), would put us on the way to explaining the fact that we can attend to our experiences, remember them, compare them, and talk about them. But thesis (3), according to which qualia have no physical effects, appears to rule out such explanations.

Yet contrary to the second part of (3), we tend to think experiences have enormous effects on our behaviour. Why listen to a piece of music if hearing it has nothing to do with whether you want to listen to it again? And consider the complicated vocabulary we use for talking about our experiences. How could we have developed it if experiences had no effects on the physical processes involved in using it? The claim that sensory experiences have really had no such effects starts to seem barely intelligible.

So have epiphenomenalists simply overlooked those facts? Not at all. As we noticed in the previous chapter, they are well aware of how violently counter-intuitive their view is. They have not taken it up from ignorance, stupidity, or perversity, but from the conviction that, when the alternatives are thoroughly investigated, they turn out to be even less acceptable. Faced by the fact that we find it hard to understand how experiences could fail to have physical effects – a really big problem for them – epiphenomenalists grit their teeth and accept that the causal inertness of experiences is just another brute fact that we have to accept, like it or lump it. If we disagree, we need good reasons, preferably in the form of arguments whose premisses cannot be denied even by epiphenomenalists.

4. How epiphenomenalists respond

It will be convenient to have a label for the group of activities mentioned in thesis (5) – noticing, attending to, thinking about, and remembering qualia. I will say they are cases of *informational contact*. Epiphenomenalists feel able to maintain that these activities are not caused by qualia because they believe informational contact is well enough explained if the activities in question are caused by the physical processes which themselves cause the qualia. When I see a meteor, for example, the light strikes my retinas and causes visual qualia. When I attend to these qualia or compare them or exclaim 'Did you see that one?' – activities I think of as caused by my experience – the epiphenomenalist will tell me that what actually caused them was not the experience or quale itself, but the underlying physical processes: the quale only appears to have been the cause. So, on their account, the common and natural assumption that informational contact is caused by the non-physical qualia themselves is to be replaced by the idea that it is caused by something else: the underlying physical processes. That looks like a clever suggestion. Does it work?

One thing quickly becomes clear. Causation by itself is not enough to get the epiphenomenalists over their big problem; they have to bring in some further condition. For if the fact that certain qualia are caused by brain processes were enough by itself to put me into informational contact with those qualia, then (since the example is arbitrary) I would be in informational contact with *anything whatever* that happened to be caused by those brain processes. For example, if they happened to cause some hairs on my head to turn green, I'd be put into informational contact with those green hairs; I'd be able to notice them, attend to them, think about them, and so on. Since that is obviously not true, mere causation cannot be enough for what epiphenomenalists offer as a substitute for real causation by qualia. If all that can be said about the non-physical item which they say constitutes the experience is *that it is caused by certain brain events*, then it need not be an experience at all, still less need I be in informational contact with it: no one need be in contact with it. After all, the brain events in question (the ones caused by my meteor experience) might cause a whole range of things other than qualia: electric currents, for example, or minute changes in the temperature of the air round my head; and very clearly I would not be in informational contact with them. So epiphenomenalists need to impose further conditions. What might they be?

5. Two counter-examples

The trouble is that epiphenomenalism has only limited resources for framing further conditions. It seems to all of us that our experiences really cause physical events, as when the flash of a meteor seems to cause me to say 'Look!'. And what epiphenomenalism needs is a substitute account which doesn't involve a non-physical quale causing a physical event such as speaking. This substitute must be plausible, yet the available resources are very meagre. They seem to be no more than (a) the physical component of the supposed epiphenomenalistic world; (b) non-physical qualia; (c) causation by the physical; (d) other relations such as similarity. Could these possibly be enough?

The most attractive suggestion for an additional condition is this: the physical cause of the experience must share some significant similarities with both the experience itself and the activities involved in informational contact with it. The idea would be that there are systematic correspondences between those activities on the one hand, and the physical events which allegedly cause the experience on the other. Relationships among the physical causes would be mirrored by relationships among the activities. In other words, there must be *isomorphisms* between the experience and the physical processes which cause it, and also between the latter and the processes involved in informational contact. When, for example, I hear the clock striking three, the distinctive sound of the chimes occurs three times, with definite intervals between them. It seems reasonable to suppose that hearing the chimes involves corresponding processes in my brain. It also seems reasonable to suggest that when I attend to the chimes, think about them, compare them with other experiences, and talk about them, the physical processes underlying these activities include components which in at least some respects reflect features of the experience itself, such as the gradual fading of each successive sound. The present suggestion on behalf of epiphenomenalism, then, is that although causation alone would not ensure that the alleged non-physical qualia were mine in the first place, still less that I was in informational contact with them, causation plus those isomorphisms would. Additionally, this account would explain why we are tempted to say that the qualia actually cause the activities involved in informational contact with them: it would explain the *appearance* of causation.

But that would still not be enough. Imagine a TV set playing inside an unoccupied room. Unknown to you a team of neurologists have connected your brain to it, and it is showing a sequence of images which exactly mirror your experiences and run parallel to events in your brain – although nobody is actually seeing these images. What is more, the neurologists have arranged for these parallel TV images to be *caused* by your brain processes. The key question now is this: do the causal links from your brain to the TV together with the resulting isomorphisms ensure that the TV images are your experiences? Do they put you into informational contact with them, so that you can notice them, attend to them, remember or compare them? Obviously not. There is no reason why you should know anything about the TV images, or even that they exist. You don't *necessarily* pick up information about them, so the mere fact that they are caused by and isomorphic to your brain processes doesn't transform them into your experiences (or anyone else's, for that matter). Still less does it enable you to notice them, attend to them, think about them, remember them, or compare them. Clearly, causation and isomorphism are not capable of providing for informational contact.

To reinforce those points here is another comparison, perhaps a bit less far-fetched. Electric currents can induce other electrical activity. Suppose, then, that the brain processes which epiphenomenalists suppose cause non-physical experiences also induce patterns of low-level electrical activity isomorphic to the original processes. There might actually be such induced currents occurring in my own head – I don't know whether there are or not – so suppose there are. In that case these cranial currents would be both caused by and isomorphic to the brain processes which according to epiphenomenalism cause my non-physical qualia. Yet since I know nothing about them, their mere occurrence would not be enough to ensure that they were my experiences, or that I could notice or attend to them, still less that I could compare or talk about them. And clearly, things that no one can notice, attend to, think about, compare, or remember, are not experiences at all.

I think those counter-examples prove that the epiphenomenalists' best response to their big problem is not good enough. Causation from the physical to the phenomenal, even when combined with isomorphism, is not an adequate substitute for causation from the phenomenal to the physical. This means that epiphenomenalism is internally inconsistent: clause (5) in its definition requires informational contact; clauses (1)–(4) rule out the possibility of informational contact. It follows that an epiphenomenalist world is not logically possible: epiphenomenalism could not possibly be true.

6. Epiphenomenalism's basic mistake

The reason epiphenomenalism gets into this mess is that it starts from a way of thinking about consciousness which, though natural, is wrong. When we think unreflectively about the nature of our experiences, we seem to be confronted by something like an internal cinema show: a succession of visual, auditory, olfactory, and other representations flowing through our minds. That conception – sometimes called the 'Cartesian Theatre' – makes it easy to assume that even if the film were to be switched off, the machinery of brains and bodies could perfectly well continue to work normally. We should have been turned into zombies, but no one would or could notice any difference.

The counter-examples illuminate the wrongness of that conception of consciousness. It is not something that could be stripped away from a person while leaving the rest of the system operating as usual. It is not a separable process – not like a jacket or paint or, indeed, like any kind of object or stuff. It is inextricable from the workings of the system, and not just from the processes contributing to the system's abilities to notice, attend to, think about, compare, and remember perceptual events. Of course, I have not so far attempted to explain how I think consciousness *can* be inextricable from those other processes; my answer will come in the following chapters.

Since the conclusion of Section 5 – an epiphenomenalist world is not logically possible – plays a crucial role in the overall argument of this chapter, the following three sections are designed to meet possible objections; they also bring in supplementary considerations. The main line of argument continues at 7§10.

7. Why do the counter-examples work?

A key consideration highlighted by the counter-examples is that when we notice, attend to, compare, or talk about our experiences, we *get information* about them. Not even epiphenomenalists could reject that view, since denying it would not only be strange and counter-intuitive, it would also be nonsensical. Suppose they said someone had noticed a meteor, or had compared two meteors, yet had not picked up any information about what they noticed, and knew nothing about it. Or suppose they said someone claimed to remember having seen a meteor but had not picked up any information about it and still knew nothing about it. Those claims would make no sense. Gathering information is an essential part of the activities in question. If you notice something, however fleetingly, you learn something about it, even if the information is incomplete or sketchy. That is even clearer for the cases of attending, thinking about, comparing, and remembering things: it makes no sense to say you could think about something you knew nothing at all about. The counter-examples show vividly how the fact that those activities depend on acquiring information rules out the possibility that physical-to-phenomenal causation, even with the required isomorphisms, should be capable of accounting for informational contact. What epiphenomenalists put forward as a substitute for phenomenal-to-physical causation turns out to be hopelessly inadequate. (There is more about information in the next chapter.)

Nor, I had better add, do epiphenomenalists have the option of simply jettisoning thesis (5), according to which we can notice, attend to, think about, compare, and remember features of our non-physical qualia. They must accept it, for it is only because they themselves can notice, attend to, and think about conscious experiences that they are in a position to theorize about them.

8. Is there a way out for epiphenomenalists?

I have argued that epiphenomenalism both requires informational contact and rules it out – because it doesn't allow for causation from the phenomenal to the physical. Epiphenomenalists might now suggest that informational contact could be provided for by causation *within* the non-physical realm: from one lot of non-physical phenomenal items to another lot; all the work of informational contact being done by the non-physical qualia themselves. But epiphenomenalism lacks the necessary resources even for this. Take the case of remembering. To remember an experience you have to be able to retrieve stored information about it; some of the information you originally picked up has been retained, and typically it has been unconscious for at least some of the time. This means that non-conscious processes are involved in storing the information – something that presumably depends on the persistence of traces left by the original experience. Further, in order to retrieve that stored information, you need to have conceptualized it: to have classified it somehow, even if only partially and loosely (perhaps as 'a bright streak in the sky'). Yet an epiphenomenalistic world consists only of the physical world as conceived of by physicalists, plus qualia – nothing else. Now, qualia are supposed to be conscious by their very nature; a person's stream of consciousness is supposed to be a stream of non-physical effects of processes in the person's brain, continually succeeding one another. This means that qualia cannot serve to store information. Being fleeting entails they cannot be the persisting traces left by earlier events. In addition, they are necessarily conscious and therefore cannot serve as the necessary unconscious memory traces. Moreover, being only the supposed non-physical effects of physical processes, they themselves can't engage in conceptualization or other cognitive activities, since these again would require them to be able to store the information that was to be conceptualized. So, given that the activities in question cannot be performed by qualia, they must, if performed at all, be performed by items in the physical component of the epiphenomenalistic world. That rules out the present suggestion.

Or is that conclusion too quick? Why shouldn't a quale which is a component of an experience cause a different kind of quale which is (for example) a memory trace? No: that is impossible. The trouble is that epiphenomenalism requires both of these non-physical items to have been caused by something physical: both are supposed to be effects of physical causes. It follows that neither could be a cause of the other. The non-physical quale could no more cause a non-physical memory trace than the film image of a trigger being pulled causes the later image of smoke from the gun's muzzle. It may seem that pulling the trigger caused the image of the smoke, but it didn't. So epiphenomenalists are compelled to hold that in spite of appearances, non-physical items don't cause other non-physical items. What at first looked like a refuge from our earlier conclusions is blocked.

Could that difficulty be overcome by revising or even abandoning thesis (3), according to which the non-physical items are wholly caused by physical processes? After all, some events seem to have two distinct causes, as when a man is hit simultaneously and fatally by two bullets, either of which would have killed him. Why shouldn't a quale cause a non-physical memory trace which was simultaneously also caused by something physical? That suggestion won't work either, because the supposed non-physical memory trace would be in essentially the same situation as the quale which supposedly caused it. For the same reason that the individual who is supposed to have the cause-quale cannot be in informational contact with it, that individual cannot be in informational contact with – in particular, cannot remember – the quale supposed to be a memory trace. Recall the earlier example of the TV in the unoccupied room, and imagine that the TV images were recorded on discs. There would be nothing to ensure that you noticed, attended to, or remembered them. Analogously, there would be nothing to ensure that I was in informational contact with any non-physical qualia that were caused by other non-physical events. So abandoning thesis (3) would not allow epiphenomenalists to escape the argument.

9. Am I just begging the question?

The epiphenomenalists say qualia have no physical effects; I claim they must have physical effects. Does that mean I am just begging the question? No. I am not saying, 'Qualia must have physical effects; you epiphenomenalists say they don't; so you are wrong'. I am saying, 'You epiphenomenalists have overlooked something. You haven't realized what it takes for a person to be in informational contact with qualia.' And I have explained (I think) what they have overlooked and how they have gone wrong. To make the key points vivid I have described two counter-examples. These illustrate how causation from the physical to the phenomenal plus isomorphism is not an adequate substitute for causation from the phenomenal to the physical. Yet the suggestion that it is an adequate substitute is epiphenomenalism's only hope because it has no resources beyond (a) physical events and processes; (b) the alleged non-physical components of experience; (c) causal relations – when the direction of causation is from the physical to the phenomenal; (d) relations such as similarity and simultaneity. I have argued that those resources are not up to the job. (An objection might have occurred to you. Couldn't epiphenomenalists appeal to a different sort of substitute for causation, on the lines of 'if x were to happen, then y would happen'? No. Such conditions would also be satisfied in the TV and cranial currents cases.)

I conclude that the epiphenomenalism represented by theses (1)–(5) is internally inconsistent. It cannot deny that we notice, attend to, think about, compare, and (on occasions) remember features of the non-physical qualia supposedly involved in having experiences. At the same time, by insisting that these qualia have no physical effects, it deprives itself of the means to explain those facts. The conclusion is that an epiphenomenalistic world is not logically possible. That completes the first part of the main argument.

Epiphenomenalism has always struck people as strange and counter-intuitive. Even its exponents have admitted that much. But although there are relatively few epiphenomenalists around today, plenty of philosophers think there *might* have been an epiphenomenalistic world. I am taking special care over both my attack on this view and possible defences of it because the assumptions underlying it are responsible for a great deal of cockeyed thinking about consciousness.

10. Zombism requires the possibility of an epiphenomenalistic world

Given the conclusion of Section 5 – an epiphenomenalist world is not logically possible – it will be fairly straightforward to show that a zombie world is not possible either. I shall argue that if a zombie world were logically possible, then so would be a world satisfying theses (1)–(5). It will follow that, since we have seen that such a world is not logically possible, neither is a zombie world. However, you don't have to be an epiphenomenalist to hold that a zombie world is at least *possible*. So, it might at first seem that you could safely endorse zombism provided you kept clear of epiphenomenalism. But that option is blocked. No matter what other views you may hold about consciousness and its place in nature, zombism commits you to the logical possibility of an epiphenomenalistic world satisfying (1)–(5).

There are four main views about the place of consciousness in nature: physicalism, idealism, panpsychism, and dualism. Only dualism even looks like a safe home for zombists, as a glance at the others will show.

According to *physicalism* nothing exists but the physical, consciousness being provided for by purely physical means. But you can't be a physicalist and a zombist, since zombism as defined entails that consciousness involves something non-physical, in which case physicalism is false. If some physicalists claim a zombie world is logically possible, they are mistaken. Physicalism is not a safe haven for zombism.

According to *idealism*, nothing exists but the mental, which is claimed to be non-physical. That obviously rules out the view that our world is physically as physicalists claim it is. But can idealists nevertheless maintain that a purely physical world is still logically possible, and might be a zombie world? No, since their whole position depends on arguments purporting to show that the world could not possibly be as it is conceived by physicalists. That means their position rules out the possibility of a zombie world as defined. Idealists, like physicalists, cannot consistently be zombists.

You may recall that according to *panpsychism* every physical object has mental components or mental properties (5§6). Panpsychists typically hold that to the extent that our world is describable in terms of physics at all, it is as physicalists say it is – and has the key feature that all physical effects have physical causes. For them, however, physics doesn't go deep enough. They hold that in spite of what they take to be the fact that consciousness cannot be captured in terms of physics, it nevertheless underlies or provides for the physically describable component of the world and for that reason may even itself be described as broadly physical. But panpsychists have to decide between two possibilities: they must either (a) allow for the logical possibility that this special consciousness-providing component of the world should be removed, leaving behind a zombie world where all effects were produced physically, or (b) rule that possibility out.

Consider alternative (a). Clearly, if the special component could be removed from the world without necessarily altering the purely physical facts, then if it were removed (leaving behind a zombie world) it could be put back again. Thus (a) would commit panpsychists to the view that a zombie world, if a suitable consciousness-providing component were *added* to it, would be transformed into a world whose human-like inhabitants were conscious. And we can easily see that that commits them to the possibility of epiphenomenalism. To start with, there is no logically necessary connection from a cause to its effect: the relation between them is contingent. As David Hume remarked, a cause and its effect are 'distinct existences', meaning roughly that the one doesn't necessarily bring the other with it. If *a* causes *b*, it is logically possible that something other than *a* should have caused *b*, and it is also logically possible that *a* should have caused something other than *b*. This doesn't mean there is no necessary connection of any kind from cause to effect. Indeed, it is generally agreed that there is a necessary connection – but the necessity is taken to be only natural, not logical. So, panpsychists who accept alternative (a) are committed to the possibility of a world which started off as a zombie world but then had a suitable non-physical consciousness-providing component added to it. Given that causation is contingent, these panpsychists are therefore also committed to the *possibility* that this special component should be caused by the physical component, without having any effects on the latter – which means they are committed to the possibility of epiphenomenalism.

The other alternative for panpsychists is (b): that a zombie world is logically impossible. So, if they choose alternative (a), they are committed to zombism and the possibility of epiphenomenalism. If they choose (b), they are not zombists at all (although this is a position that panpsychists don't actually seem to hold). Those who take the first alternative are vulnerable to the argument of §§4–6 above. And if any of them take the second alternative, they can be ignored from our point of view.

So physicalists and idealists cannot consistently be zombists, while panpsychists who are also zombists must accept the possibility of epiphenomenalism. That leaves *dualism,* the fourth main view about the place of consciousness. It comes in three varieties: epiphenomenalism, parallelism, and interactionism. No need to argue that epiphenomenalists are committed to the possibility of an epiphenomenalistic world, of course. And given the contingency of causal relations, parallelists too are committed to that possibility. In spite of their view that there is no causal interaction between the physical and the mental, they are bound to concede it is at least logically possible that there should have been. In other words, they must agree it is logically possible that epiphenomenalism should have been true.

Just one main variety of dualism is left – interactionism. Is it, at last, a safe haven for zombists who resist epiphenomenalism?

11. Interactionism won't get zombists off the hook

According to interactionist dualism, what provides for consciousness is a special, supposedly non-physical, component of the world. This non-physical component and the physical component each have effects on the other. You might at first guess that this would allow interactionist zombists to escape commitment to the possibility of epiphenomenalism. Rejecting the physicalists' account of our world, they hold that some physical events are caused non-physically. If that were actually the case and this were an interactionist world, then stripping consciousness out of it would not make a zombie world. It would make a world where what was left of the human race was incapable of normal human behaviour because the non-physical causes of that behaviour would have vanished. Without the desires and feelings that cause normal behaviour, what was left of us would lie about mindlessly twitching, as our autonomic nervous systems responded to external stimuli. However, that consideration doesn't let interactionists off the hook because only *zombist* interactionism is relevant – and it requires at least the possibility of a zombie world. Here, then, is a simple argument to show that zombist interactionists, even though they deny that epiphenomenalism is actually true, must accept that an epiphenomenalistic world is possible.

The argument starts by supposing a zombie world exists. Now, according to interactionism there is a special non-physical something which ensures there is consciousness in our own world, so we next suppose that this supposed zombie world is modified by the addition of just such a consciousness-producing item. And here is the decisive step. Since causal relations are contingent, it must be logically possible that this non-physical something is caused by physical processes inside the brains of the zombie world's inhabitants, yet does not itself cause any physical events. This means that the result of adding it to the zombie world in that way would be an epiphenomenalistic world, if such a thing were possible. It follows that zombist interactionists, like it or not, are committed to the view that an epiphenomenalistic world is logically possible. (I had better note that there is a range of possible dualist interactionist views, from the idea of a single centre of consciousness in the universe, through Cartesian dualism, to epiphenomenalism and parallelism. But the argument does not require the special non-physical 'something' to conform to any particular model; it depends only on the fact that all varieties of interactionism postulate a non-physical source of consciousness.)

Summarizing:

(1) According to interactionism, there is a special non-physical something which provides for consciousness.

(2) Because causal relations are contingent rather than logically necessary, zombist interactionists must accept that this special non-physical something could be added to a zombie world in such a way that although it was caused by physical things in that world, it had no physical effects.

(3) The result would satisfy conditions (1)–(5) for an epiphenomenalistic world.

(4) So, zombist interactionists are committed to the logical possibility of an epiphenomenalistic world.

I argued first that an epiphenomenalistic world is not logically possible. I have just argued that if a zombie world were logically possible, an epiphenomenalistic world would be logically possible too. It follows that if both parts of the argument are sound, then *a zombie world is not logically possible.*

12. Conclusion

I have offered a fair amount of detailed argument in this chapter because its conclusion is highly significant. If zombies had been logically possible, then something non-physical would have been needed in order to provide for consciousness – and the attempt to explain it would have crashed on the rocks of brute inexplicable fact. If the argument is sound, though, there is no serious objection to physicalism. Assuming neurophysiology doesn't reveal hitherto undiscovered gaps, the purely physical facts about us are sufficient to account for consciousness.

We had better acknowledge, however, that it doesn't follow that physicalism is true. I know of no reason why the world could not have turned out to be a dualistic one, provided it were interactionistic. Still, the thought that zombie worlds are possible has been a major obstacle to the physicalist project. Removing it makes that project much more promising than some people continue to believe. There can no longer be serious philosophical objections to physicalism, and dualism no longer seems worth bothering about.

We still face a huge question. Even though the physical facts entail the facts about consciousness, how can that be explained? I have argued that consciousness is not the sort of thing that could be stripped away from a person, and is somehow inextricably involved in activities such as noticing, attending to, comparing, and remembering things. But the mental and the physical continue to seem like two very different kinds of thing, and it can seem impossible to bridge the gap between them.

8

The basic package

1. Starting with oysters

This is the book's turning point. So far I have been explaining the main philosophical problems of consciousness and discussing the theories they have provoked. From now on I shall present and defend what I believe is a sound approach to solving them: a version of functionalism.

Discussing questions about robots and zombies has helped to clarify the issues. The notion of zombies has a hold on many philosophers, but it ought not to. It arises from assumptions which on investigation turn out to be contradictory: such things are impossible. Robots however are not just possible: thanks to the flourishing of robotics we are surrounded by these artefacts in large numbers and great variety (think of driverless cars, Mars rovers, car assemblers, systems that play go) and they raise further problems. It has been useful to ask whether robots can be genuinely intelligent or genuinely conscious. None of the arguments against those possibilities – including Searle's famous and still influential Chinese Room – has stood up to close examination (although, of course, further investigation might yet reveal a fundamental objection to robotic intelligence or consciousness). But evidently, we cannot make real progress towards solving our philosophical problems without going more deeply into what it takes to be conscious.

We can tell we are conscious because we *are* conscious. But when we try to discover whether other kinds of systems are conscious we need a different approach; we have to rely on their behaviour and what goes on inside their bodies. Examples such as the Giant and the Block machine strongly suggest that the mere fact that a system behaves appropriately does not guarantee it is genuinely conscious; it looks as if philosophical behaviourism fails. But suppose we are considering a certain system and already know both how it behaves and what goes on inside it. Could that deliver a guarantee of consciousness? The question is hard because, even though consciousness doesn't require a special non-physical component, we are still a long way from knowing what it takes for *any* system to be conscious.

Consider these remarks of John Locke's:

> *Perception*, I believe, is, in some degree, *in all sorts of Animals*; ... We may, I think, from the Make of an *Oyster*, or *Cockle*, reasonably conclude, that it has not so many, nor so quick Senses, as a Man, or several other Animals; ... – But yet, I cannot but think, there is some small dull Perception, whereby they are distinguished from perfect Insensibility. (*Essay,* 1689/1975, II.ix.12)

It is worth reflecting on those words for two reasons. One is that a good strategy for gaining an understanding of phenomenal consciousness in general is via *perceptual* consciousness. Perceptual consciousness is after all what most of us enjoy during our waking hours, and if we can understand perceptual consciousness, then it seems likely that other varieties of phenomenal consciousness, for example dreaming and daydreaming, will be explicable on that basis. The other reason is that it is a good idea to start with relatively simple organisms, and for that matter simple artefacts. Considering simple systems ought to help us to avoid being distracted or even misled by the complexities of normal human consciousness.

Locke's remarks about the 'make' of oysters and cockles seem to imply he was acquainted with their internal workings. If he really did know how they worked, then no doubt that would have helped him to decide whether or not they had 'some small dull perception'. But considering the state of biology in his day, we can reasonably guess he did not have much knowledge of that sort, and that the main determinant of his thinking about bivalves was not their innards – which would not have been very informative to seventeenth-century eyes – but their behaviour. Now, when we describe creatures' behaviour we do not usually put it in terms of the movements of their bodily parts ('right hind leg slowly raised', 'eyes move quickly to the right', and so on). Typically, we describe it in terms that could be counted as broadly psychological, and might also be appropriate for describing human behaviour: the dog was 'looking for the ball', the chimp 'tried to get at the ants'. Descriptions like these imply the creature has something like mental states, even if they are primitive when compared with our own. In fact it seems impossible that anything should be perceptually conscious unless it had mental states; consciousness seems to be inextricably linked with intelligence (a point I will develop later). In this chapter I will argue first that nothing could be perceptually conscious unless it had at least something like mental states, and then that there is a certain set of capacities all or most of which are at any rate necessary for perceptual consciousness.

2. Perception and pure reflex systems

We know human consciousness is more than a mere parade of experiences. Notably it involves the capacities to notice, attend to, and on occasions remember and compare experiences. But some of those capacities are quite sophisticated. (Do oysters compare their experiences? Not likely.) We need some idea of the *minimum* requirements for perceptual consciousness.

Suppose I am looking at a book. By seeing it I learn such things as its position, shape, and colour; analogously for perception by the other senses. Quite generally, perception involves learning: to perceive is to acquire information about what is perceived. So, to understand what is involved in conscious perception we need among other things to understand what is involved in acquiring information.

It will be useful to start by focusing on a certain very broad class of systems which may appear to acquire information, but actually don't because they simply react in fixed ways to external stimulation. When you press a certain key on a piano it sounds middle C; when you put a certain weight of sugar on a scale, it reads '1 kilo'. Examples mentioned by Locke included sensitive plants and 'the turning of a wild Oat-beard by the Insinuation of the Particles of Moisture' (*Essay* II.ix.11). In general, the behavioural repertoires of such systems fit the simple S-R pattern we have already noted: each of a fixed set of possible stimuli causes a certain fixed response. So this type of system can be represented by a list of possible stimuli and the fixed response to each:

S1 causes R1
S2 causes R2
...
Sn causes Rn.

I will call systems whose behaviour can be represented in that way 'pure reflex systems'. Simple though this scheme is, a pure reflex system's behaviour can seem intelligent, and would enable it to survive in a suitable environment. Pretend for example that a certain kind of fly were a pure reflex system. (Real flies are not pure reflex systems: even the fruit fly learns to find its way home, acquiring new ways of reacting to its surroundings.) And suppose for argument's sake that this pretend fly's entire behavioural repertoire could be captured by a set of statements such as these:

Lack of contact of its feet with a surface *causes it* to buzz and fly about.
Contact of its feet with a surface *causes it* to stop buzzing and walk about.
Detecting certain chemicals typically produced by rotten meat *causes it* to fly or crawl in ways that maximize such detection.
Proximity to a source of such chemicals *causes* its mouth parts to go into action.

A fly with those fixed responses to the given stimuli plus a few others, notably to do with reproduction, will have a good chance of surviving in the environment where it has evolved. It is so constructed that a given stimulus causes it to respond in a way that typically makes it liable to receive a different stimulus; its response to that stimulus in turn puts it in the way of receiving a third kind; and so on throughout its existence. Given a reasonably stable environment, the creature may live a perfectly coherent life. Merely behaving along those fixed lines will result in behaviour that would be construed as intelligent if we didn't know how the creature worked. (Might some of us fit that pattern?)

However, this creature's responses are hard-wired, and for that reason alone it doesn't really learn. Genuine learning – really acquiring information – would require its behaviour to be *modified* as a result of encounters with its environment – which is ruled out by the definition of pure reflex systems: their responses are fixed. So, since perception requires the capacity to learn, pure reflex systems cannot perceive either – or not in the relevant sense. It may still be natural to say that our stimulus–response fly 'perceives' the rotten meat. But perception of that kind doesn't concern us. If we allow the word to apply in such cases, then we had better say that the present point is that the relevant type of perception – the kind needed for perceptual consciousness – requires the capacity to learn.

However, we need to be careful what we count as learning. Although the learning involved in perception requires the system's behaviour to be modified as a result of encounters with its environment, it is easy to think up cases where that condition is satisfied but the system still doesn't really learn. Suppose the 'make' of a certain pure reflex system is such that exposure to certain sequences of events causes its original reflexes to be replaced by new ones. A particular kind of butterfly, for example, might start its life with a built-in set of responses to certain fixed stimuli. These fixed responses might include the triggering of a certain type of zigzag downward flight when the butterfly's retinas are hit by patterns of light from an approaching swallow: a fixed response usually allowing the butterfly to escape. However, chemicals from the nectar of a particular kind of flower might cause changes to the insect's internal structures, resulting in the extinction of that particular response. In that case its behaviour would indeed have been modified as a result of encounters with things in its environment, but it would not have learnt anything – any more than a piano would have learnt something if its tuning had been thrown out by heat from a fire.

We can make progress if we reflect that in order to be sensibly described as acquiring information, the system must be able to use it. The imagined butterfly's behaviour is modified all right, but there is no sense in which it acquires information it can use. And clearly, a system cannot be said to use information unless it can to some extent guide or control its behaviour in the light of that information.

3. Perception and control

Unlike pure reflex systems, genuine learners – acquirers of information that they can use – control their own behaviour. That is, on the basis of their assessment of what is happening around them, they monitor what they are doing and modify it when that seems appropriate given their goals. Perception in the full sense enables perceivers to do just that: guide their behaviour in the light of what they learn about what is going on: in particular, about the effects of their behaviour on their environments.

We have reached a watershed. The difference between pure reflex systems and systems capable of such learning is enormously significant. I am not saying there a sharp division between pure reflex systems and ones capable of controlling their own behaviour. On the contrary, there is plenty of room for systems which are neither, a whole spectrum of cases (the Giant is just one). But it is still very useful to keep in mind this contrast between pure reflex systems and genuine learners; it helps to highlight the main features of the latter.

How systems capable of controlling their own behaviour may have evolved from simpler ones is something I leave to your background knowledge or imagination. What we need to do now is consider what *capacities* these significantly more sophisticated systems must have. Monitoring and modifying behaviour in the light of incoming perceptual information are evidently not simple processes: they involve a whole range of lower-level capacities. On reflection, it is a striking fact that several of these lower-level capacities come in a package: virtually inseparable.

Our cat Zoë is chasing a mouse. This behaviour might be a mere fixed response to a stimulus – perhaps to hearing a squeak. However, amateur observation of cats over many years has convinced me that although these animals have strongly built-in patterns of behaviour when certain stimuli impinge on them, a lot of what they do doesn't fit the simple stimulus-response pattern. Indeed, it is not too much of a stretch to call them *deciders* in the sense that they monitor and control their own behaviour on the basis of incoming perceptual information. I will therefore assume that when Zoë is engaged in chasing a mouse, she has – however fleetingly – the *goal* of catching the mouse. She may be distracted by a barking dog, and if the dog is close enough she may abandon the chase and climb a tree; if the dog is further away, she may continue her pursuit. So, she can *initiate and modify* her behaviour, given her goal. In addition, she can be said to *choose* between those alternative courses of behaviour: escaping the dog by climbing the tree, or continuing to chase the mouse (a pretty basic sort of choosing: see below). Also, if her behaviour and its modifications are to have a chance of achieving her goal, the incoming information – which in the case of vision starts from the impact of photons on her retinas – requires some kind of *interpretation*. If she hears a squeak, she may interpret it as merely *human-generated noise*; on the other hand, she may interpret it as *mouse-caused* (or whatever the equivalent feline conceptualization might be: see later). Such interpretation helps to ensure that the incoming information enables her to *assess* her situation. But of course, if the interpreted information is to be of any use, she must be capable of *storing and retrieving* at least some of it. Another thing that requires her to be capable of storing and retrieving information is the fact that her goal *persists*, if only for a short time.

To summarize these main lower-level capacities: a system capable of controlling its own behaviour must:

(i) *acquire, store, and retrieve* information;

(ii) *initiate and modify its behaviour* on the basis of incoming perceptual information;

(iii) *interpret* information;

(iv) *assess* its situation;

(v) *choose* between alternative courses of action;

(vi) *have goals*.

I call this the *basic package,* and say that any system with this package is a decider – regardless of whether it is a natural system like one of us, or artificial, like a robot.

Two things may cause raised eyebrows: the idea that deciders must be able to choose, and the notion of *information*. I will deal with the second in the next section; the first straight away. What is involved in choosing between alternative courses of action? I have used a cat as an example; but does that mean I think cats have free will? If not, does it follow that they can't be deciders after all? The question is puzzling. Find a cat to observe, and reflect on its behaviour. Is it appropriate to say it has free will? Or that it doesn't have free will? Hard to say, I find. Sometimes Zoë behaves as if she is simply wired to react to a stimulus like a simple S-R system, as when her jaws start snapping on sight of a bird. Sometimes, though, she seems to make a choice – as when, after spotting a bird on a branch and looking from bird to tree-trunk and back again several times, she eventually starts to climb. But I don't think we need to debate whether this sort of behaviour depends on free will (whatever that may be). I said earlier that Zoë 'can be said to choose' – and I suggest that is true regardless of one's detailed position in the interminable debate about free will. It is a fact that normal cats have the capacity to process perceptual and other information in ways that make it natural to say they choose between alternative courses of action. That doesn't automatically imply we think they have free will – if only because it is not clear what that means. We can say confidently that sticks and stones don't have free will, and no doubt the same goes for bacteria. Creatures like cats and dogs are different. However, as remarked earlier, a good reason not to pursue this particular line of thought is that there are no plausible arguments showing that free will is necessary for intelligence or consciousness.

Typically we are content to use psychological descriptions on the basis of the subject's verbal and other behaviour. In the case of languageless animals, we tend to use them by analogy with our own cases – a procedure obviously liable to error, as robot examples have made clear. The best I can do is to say that a system chooses if, all things considered – that includes its internal workings – it has the other capacities in the basic package and behaves as if, on occasions, its behaviour is influenced by taking account of a range of alternatives (see also 4§1). The sea defences of Rotterdam provide a good illustration. When serious floods threaten, an enormous barrier is raised. That involves great losses for the inshore fishing industry; on the other hand, if it's not raised and the land is flooded, there are other great losses. I understand that in order to prevent the decision to raise the barrier from being influenced by the personal and other biases of a committee of human beings, it has been entrusted to complicated software, which takes into account a vast mass of information about tides, winds, water levels, and so on. Now, this system doesn't seem to have the other capacities in the basic package; but if it had had them, I should have said it was a decider. And the fact that it works deterministically is irrelevant.

4. Information

I have been assuming we already have a working conception of information, vague though it is, and I am hoping that what has been said so far helps to explain how I understand the word as it applies to deciders. The discussion of computers in Chapter 3 noted some of the ways in which information may be processed; but further points must be spelt out.

Clearly, the *acquisition* of perceptual information by a decider consists in its internal state being changed in certain ways by the impact of the outside world on its sense organs. *Storing* such information consists in these or related changes enduring over time. It can *retrieve* stored information in the sense that these enduring changes contribute to guiding its behaviour in ways that reflect their relations to their external causes. But you may wonder what sorts of changes are involved.

Information in computers, when translated into the system's machine code, is stored in the form of something like sentences, and you might guess the same would go for deciders in general. That is hardly possible, however. One reason is the enormous volume of information acquired via ordinary conscious perception. When I look out of the window I see a brick wall. But the sentence 'there is a brick wall' summarizes only a tiny fragment of the total mass of information that I also acquire. I see, for example, that these bricks are in a range of different reddish colours; that some are worn away in different ways; I also see precisely where and how they are worn; I see that mortar is missing in various places, and again where and how; and so on ad indefinitum. The idea that each of the components of this mass of information comes packaged in anything like the form of a separate sentence seems absurd (notwithstanding Daniel Dennett's suggestion that 'seeing is rather like reading a novel at breakneck speed' (1969: 139)).

What about information in the brain? We know it contains something like 100 billion neurones, each variously interconnected with others. The storage of information in this almost unimaginably huge and complex system is determined, it seems, by the relative strengths of the connections from neurone to neurone (from the dendrites of one neurone to the axons of others). Although some philosophers argue that nevertheless much of what is going on in the brain is in fact the processing of something like sentences (Fodor 1975), I don't find the arguments for that view persuasive. In any case, even that theory requires the processes involved to consist at some level of the states and interactions of neurones and neurotransmitters and inhibitors. Recalling what was said earlier about different levels of description and explanation, those neural states and interactions can be said to occur at a *low level*. Although it is higher than the level of atoms and electrons, it is considerably lower than that at which it is appropriate to talk of 'interpretation', 'assessment', 'decision-making', and the other capacities in the basic package. Nevertheless, I take those higher levels of description and explanation to apply to the very same processes as are also describable in terms of the states and interactions of neurones and neurotransmitters (and, much lower still, in terms of the states and interactions of atoms and electrons). To talk in terms of interpretation, assessment, and so on, is not to appeal to something additional to those low-level processes: it is to *redescribe* them (more later).

It is becoming clear that we can only gesture towards answering the question of what sorts of changes are involved in the acquisition, storage, and retrieval of information in a decider. Certainly, both brains and computers provide examples; but there seems no limit to the sorts of low-level processes that could do what is necessary: no limit to other possible implementations – when what is necessary can be stated only vaguely, as above: a system acquires information when its internal state is changed in certain ways by the impact of the outside world on its sense receptors; it stores information when such changes endure over time; it retrieves stored information when these changes contribute to guiding its behaviour in ways that reflect their relations to their external causes.

The information must be, in a certain sense, *for* the system, given it has the basic package. To explain: if you load the contents of an encyclopaedia into a computer, *you* can use that stored information. But the machine itself could not use it, or not without running a special program. That information would not be for the computer itself. Similarly, cameras acquire information that is useful to us, but not to them. But you may still ask, what is required for the information acquired by a system to be for it? I hope it will become clear as we proceed that what is both necessary and sufficient is that the system has the basic package.

5. Interdependent capacities

The components of the basic package appear to be interdependent. The system cannot acquire information unless it can store and retrieve it. It cannot use stored information except by acting, which means it must have goals or aims or objectives however minimal (which also have to be stored). It cannot aim at goals unless it can initiate and control its own behaviour. It cannot do that unless it can assess its situation. That in turn it cannot do unless it can interpret incoming perceptual information. And it cannot commit to action unless it can choose or decide between alternative courses of action on the basis of its interpretation and assessment of its situation. By similar reasoning we can start from any of the other capacities and end up with the rest. So it seems these capacities are interdependent.

You may be able to think up apparent counter-examples, however. One suggestion is that the storing and retrieval of information doesn't require the other capacities. Computers store and retrieve information, after all, but their software doesn't generally enable them to interpret it or to assess their situation (as usual there are exceptions), still less does it give them goals. But that objection overlooks the point just emphasized: capacity (ii) (to initiate and modify behaviour on the basis of incoming perceptual information) concerns information that the system itself can use. In the relevant sense, this means that the system must be able to guide its behaviour on the basis of the information. In contrast, as noted, the information stored in a computer is not generally available for the computer itself to use.

There are however *degrees* to which it is appropriate to describe a system as capable of interpreting perceptual information, or assessing stored information. Perhaps the same is true of the other capacities: they don't seem to be all-or-nothing. That makes it easier to find apparent counter-examples. You might suggest that people with extreme dementia, for example, have no memory, yet still seem capable of having conscious perceptual experiences. But it is important that the sort of memory required by the basic package can be quite limited. If dementia patients are still capable of passages of coherent behaviour, then they must be able to store and retrieve some information, if only briefly, otherwise it is hard to see how they could ever set out to do anything. If they are so unfortunate as not to be able to behave coherently even for short periods, there must be a question as to whether they are still perceiving rather than just behaving automatically on the lines of a reflex system.

Another apparent counter-example is that of paralysed people. We have excellent reasons to think they are consciously perceiving the people and things around them, but if the 'initiation of behaviour' envisaged for capacity (ii) were taken to refer exclusively to overt bodily behaviour, then they wouldn't have it. Two considerations weaken the force of that objection. One arises from the fact that there are central and standard cases of perception, and also peripheral and non-standard ones. In the central and standard cases the organism can control its overt or bodily behaviour. Now, paralysed people are not central cases because they don't have that capacity; at the same time they closely resemble central cases in other respects. Their lack of the capacity in question is due to some defect or malfunctioning. (After all, any of us can be temporarily paralysed by drugs or otherwise.) The other consideration is that not all behaviour is overt and bodily. Paralysed people can at least initiate and control chains of thought, so in that sense they have capacity (ii). If they could not initiate and control any of their thoughts, it must be problematic whether they had any of the other capacities either. In any case we can take capacity (ii) to refer to all kinds of behaviour, not just overt.

Since we are talking about ways of classifying a wide range of natural and artificial systems, it would be silly to claim there was a sharp boundary between deciders and non-deciders; there will be indefinitely many borderline cases – as well as a range of different borders. But if we focus on typical systems capable of controlling their behaviour on the basis of incoming perceptual information – systems such as the cat Zoë – it seems they have all the capacities on the list (i)–(vi). These are central, standard cases. In spite of worries about possible counter-examples, it is significant that there are plenty of these standard cases. For one thing, it is typically on the basis of our encounters with such cases – in particular with one another – that we are able to learn how to apply the concepts which describe the capacities in the basic package. Normally we just don't come across systems which have only one or two of those capacities without the rest; and indeed it seems that our ability to use the concepts in question depends largely on the fact that these capacities are interdependent.

A different worry is that parts of the definition of the basic package may seem not to be applicable unless the system in question is already known to be conscious. But that need not worry us, since in the present context any such implication is to be cancelled. Take, for example, the capacity to make choices. Some people assume that only a conscious system can make genuine choices. They may be right, but the present point is that consciousness must not be made a precondition when deciding whether a given system has the basic package. What determines whether or not a system can make choices, or satisfies other parts of the definition, is for present purposes facts about its behaviour and dispositions, and about relations among these and its internal states – facts which can be established without first deciding whether or not the system is conscious.

Now, I claim the basic package is necessary for conscious perceptual experience. Is it also sufficient?

6. Concepts and language

The basic package requires the system to have something like concepts. When Zoë interprets a sound as *mouse-caused*, for example, she is in her own way conceptualizing it. Assessing her situation with respect to the barking dog, she thinks of the dog as too far away: more conceptualization. Further, in order to be able to choose between possible courses of action she must represent those possible actions in some way, which again is a matter of bringing them under concepts. Now, quite a few philosophers maintain that concepts require language. Since infants, cats, dogs, and other creatures which many of us think are conscious don't have language, then if those philosophers are right, my suggestions are wrong.

But what is involved in having concepts? You might at first think, as Socrates seems to have done, that it would require knowing necessary and sufficient conditions for something to fall under the concept in question: a definition. (A plausible example of such a definition might be, 'Bachelors are unmarried men'.) So if Socrates' assumption that concept-possession depends on knowing necessary and sufficient conditions were correct, we could hardly credit Zoë with having the concept *mouse*. She is obviously very far from knowing necessary and sufficient conditions for mousehood (murinity?). However, Socrates was surely mistaken. I myself have the concept *mouse* but I don't know necessary and sufficient conditions for mousehood; it seems to be a different sort of concept from *bachelor*. In fact I doubt whether there are such conditions, in which case, by Socrates' criterion, no one – human or not – has the concept *mouse*. No one seems to have been able to find a satisfactory set of conditions for such concepts, and in the last half-century or so the suggestion has been widely rejected by philosophers for that and other reasons.

Another suggestion is that – at least for the case of perceptible things such as mice – having a given concept is a matter of being able to recognize instances of it, or to discriminate them from other things. But that can't be right either. Hilary Putnam (1975) gave what seems like a good counter-example. Even though he can't recognize instances of *elm* as distinct from those of *beech*, he still understands the words, and in that sense has the concepts, because he belongs to a community some of whose members have the necessary expertise. This is 'the division of linguistic labour'. (Obviously, possession of theoretical scientific concepts such as *atom, weak force, Higgs boson,* and others cannot require the ability to recognize or discriminate instances, but Putnam's examples are more straightforward.)

So what about the idea that having a concept such as *mouse* requires being able to use the word 'mouse' or its synonyms correctly? If languageless animals don't have concepts, then they don't have the basic package, and if they don't have that, then according to me they can't be genuinely conscious. However, the people who urge that language is necessary for concept-possession tend to have a rather special conception of concepts. They are willing to allow that animals may have something *like* concepts and, relatedly, something like beliefs ('proto-beliefs' is one author's word). In any case the behaviour of babies, cats, and many other creatures shows they tend to group together different items in their environments, to treat the grouped items similarly, and generally to organize their behaviour in ways that come close to conceptualization, if only in rudimentary forms. The cat does actually distinguish between mice and other things, and remains motivated to catch a mouse once spotted; chimps and the famous New Caledonian crows construct tools for extracting food from otherwise inaccessible places. These creatures surely have as much as the basic package requires in the way of conceptualization. If that is right, then language is not necessary for the basic package, in which case the reasoning which threatened to force me to concede that languageless creatures can't be conscious is blocked. (Another consequence worth mentioning is that we cannot generally render the thoughts and beliefs of languageless creatures accurately. 'Human-generated noise' and 'mouse-caused' can hardly be accurate representations of Zoë's conceptualizations, since she surely doesn't have our concepts *human*, *generate*, *noise*, or even *mouse*. But that doesn't mean that felines don't have their own simpler concepts.)

Some people assume that concept-possession is all-or-nothing: an individual either does or does not have a given concept. But having a concept doesn't seem to be a simple unanalysable matter. In general – certainly in the case of everyday concepts – concept-possession involves a whole range of capacities and expectations, coming in degrees. Consider the concept *rabbit*. Various capacities and states of mind are associated with it, for example: (i) being able to use the word 'rabbit' and react appropriately when it is used; (ii) being able to discriminate rabbits from other things in normal situations; (iii) recognizing rabbits *as* rabbits; (iv) treating rabbits in ways appropriate to the individual who has the concept; (v) having certain beliefs, such as where rabbits typically live and how they behave; (vi) expecting rabbits to be furry, with long ears, prominent front teeth, and short white tails. English speakers typically have all of (i)–(v). But I have given reasons why (i) is not necessary, and there are also reasons why none of (ii)–(v) is necessary either, although it seems doubtful that a person could be counted as having the concept if they had none of these capacities and expectations. (Since the concept *concept* is very vague, there seems no reason why we should require more precision over what counts as possessing a given concept than those rough indications.)

Some creatures' behaviour may at first suggest they possess quite sophisticated concepts, yet they may still be only pure reflex systems. Salinity in the water causes oyster larvae to swim, but we don't have to say these creatures have the concept *salinity*; their behaviour is merely reflexive. Ticks, which also might be pure reflex animals, have a complicated receptor system in 'Haller's organ', located in their legs. This organ 'receives external stimuli in terms of temperature, humidity, carbon dioxide concentration, ammonia, aromatic chemicals, pheromones, and air-borne vibrations' (Hillyard 1996: 20). In spite of the sophistication of the concepts in terms of which scientists characterize the stimuli discerned by these creatures, we can hardly say they themselves are sophisticated, or possess any concepts at all.

So, what is needed for genuine concept-possession? The Giant and other counter-examples imply that philosophical behaviourism is mistaken, so that even behavioural dispositions like ours would not be enough. It is reasonable to suggest that concept-possession requires the sorts of psychological states and capacities that were mentioned above in connection with *rabbit:* being able to discriminate rabbits from other things; recognizing them as rabbits; treating them appropriately; having certain beliefs about rabbits; expecting them to conform to certain descriptions. Those capacities and states depend on more general capacities to acquire, store, and retrieve information; to initiate and modify behaviour on the basis of incoming perceptual information; to interpret information; to assess one's situation; to choose between alternative courses of action, and to have goals. In other words, they depend on the basic package.

I suggest that those capacities amount jointly, if not individually, to what is necessary and sufficient for having a minimal kind of psychology, and thus (as I understand these vague but still useful everyday notions) a minimal kind of intelligence. My proposal is that to possess concepts is, minimally, to be a system with the basic package – a decider. (This cannot amount to a definition since the definition of the basic package itself appeals to conceptualization.) Possession of a given concept will then be a matter of having a sufficient number of the sorts of interrelated psychological states and capacities mentioned in connection with *rabbit*.

If I am right in thinking that perceptual consciousness requires the basic package, then consciousness requires that minimal kind of intelligence.

7. Do robots have the basic package?

I take it, then, that what I have been calling 'intelligence' can for our purposes be the minimal psychology constituted by the basic package. My own view is that lots of languageless creatures have the basic package; they are intelligent in the sense relevant to the question whether robots are intelligent. We know that some robots with the right behavioural dispositions do not have the basic package and are therefore not intelligent in the relevant sense. One interesting example was the generalized version of Block's machine. Block's reasons for rejecting its claim to intelligence are also reasons for holding that (according to the moderate realism of everyday psychology (4§7)) it would not have the basic package; in particular, it could not really interpret information, assess its situation, or choose between alternative courses of action. I have argued that nevertheless the usual objections to intelligence in robots do not work; we have seen no good reason to suppose that *no* robot could be intelligent. A robot that *would* have the basic package because it would have both the right dispositions and the right kinds of internal functions is the 'computers-for-neurones' system to be considered later (11§6).

8. The machine table robot

It will be worth taking a little time to think about another imaginary kind of robot. It certainly has the right dispositions, and it might at first appear also to have the right sorts of internal processes for the basic package. Discussing it will help us gain a better grasp of the conditions necessary for possession of the basic package. To give you a very rough preliminary idea, the workings of a given human brain – say my own – could in principle be represented by a machine table (specifying, for each possible internal state of the system and each possible input to it, which state and output come next: 3§2). This machine table is then connected to a human-like artificial body. Given certain not unreasonable assumptions, the resulting robot would behave exactly like the person whose brain its machine table represented. However, there are reasons to reject the view that it would have the basic package. (If what follows threatens to become a little too technical for your taste, skip to the last paragraph of this section.)

The machine table robot is superficially like a normal human being, but its behaviour is controlled not by a brain, but by a machine table of the kind explained at 3§3. This machine table is modelled on a normal human brain: my own, which for the sake of argument I will assume is normal. We start from the fact that at any instant my brain receives a complicated pattern of inputs from all sense organs at once. We can now think of all possible total patterns of sensory inputs, making the admittedly very large assumption that time can be split into sufficiently brief instants to ensure that although there is a vast number of these total patterns, there are still only finitely many. Consider next that each total pattern of sensory input at an instant reaches the brain when it is in one of a vast number of possible states, and that typically a given total pattern of instantaneous sensory input will cause a change in the brain from one state to a different one.

Often, if not always, such a change of state will cause a certain total pattern of output to the motor nerves. So there will be a vast but finite number of possible total patterns of instantaneous sensory inputs to the brain, a vast but finite number of possible total brain states, and a vast but finite number of possible total patterns of output. Here, then, are the key moves in the description of this robot. Each possible input, possible state, and possible output is assigned an arbitrarily chosen *number*. It follows that on our assumptions, the workings of my brain can be represented by a monstrously large machine table, consisting of a set of sequences of four numbers. In each of these sequences the first number represents a possible input, the second and third each represents a possible state, and the fourth represents a possible output.

So the idea of this machine table for the Kirk-brain-representing system is that, for each of the possible states it might be in, its input when in that state fixes what its next state and its output will be. A single sequence from the machine table would have the form 'w, x, y, z', meaning: 'If the number w is put in when the system is in state numbered x, it goes to the state numbered y and puts out number z.' Note that all the possible total instantaneous patterns of sensory input to my brain and all my possible total brain states would be represented by numbers in the sequences belonging to this machine table, ensuring that no matter what inputs the environment might have thrown at my brain, every possible effect on my brain was taken into account. It follows, given our assumptions, that if my brain were to be replaced by a computer programmed in accordance with this machine table, the resulting Kirk-brain-representing system would have exactly the same behavioural dispositions as I have. Imagine now that my body is simulated by a metal-and-plastic system (no worrying surgical operations). The result will be a robot with exactly my behavioural dispositions.

Does this robot have the basic package? It certainly behaves as if it did, since it behaves like me. Unlike the Giant, it is not a kind of puppet. Unlike Block's machines, it does not have all possible life histories programmed into it. Nevertheless, its internal workings just don't conform to the moderate realism of everyday psychology. As we saw earlier, everyday psychology is not purely behaviouristic, but demands moderate realism about the nature of our internal processes, and doesn't allow wants, beliefs, and the rest to consist purely of behavioural dispositions. I cannot say precisely what it requires in addition, since it is far from a precisely defined theory. However, I assume it requires at least that certain kinds of *causal* relations hold between our various experiences, wants, beliefs and inclinations. I happen to love the smell of roasting coffee, and smelling it might easily cause me to say 'I love that smell' and at the same time to acquire the belief that there is roasting coffee somewhere around. Since our robot behaves just like me, it will produce that same utterance in the same situation. To those ignorant of its internal workings it will seem to have mental states standing in the same causal relations as mine. But we can see it has no such states, and therefore that there are no such causal relations – because at any instant its *total* state is represented by a single number, and for any extended period of time its successive states are still represented only by a succession of single numbers. Its total state would have to represent not only its detecting the smell of coffee but also its holding a whole lot of beliefs and wants, including the belief that 2 + 2 = 4 and the desire to drink coffee. Therefore nothing is going on inside it that might constitute its remark being caused by smelling coffee rather than by, for example, believing that 2 + 2 = 4. That and similar considerations make clear that this robot's internal processes are very far indeed from conforming to the moderate realism of everyday psychology. It doesn't have the basic package in spite of appearances.

You might think the fact that the machine table robot has the right dispositions undermines the claim that consciousness gives creatures an evolutionary advantage (1§2). But that would only be a worry if the chances of its having evolved naturally were comparable with the chances of something else with the same dispositions having evolved naturally – and in view of this machine's internal structure, those chances are clearly not comparable. So, I see no problem here.

None of that gives us any reason to conclude that no possible robot could have the basic package. It shows only that some systems whose internal workings might appear to let them count as deciders may fail to qualify. It illustrates one of the ways in which the moderate realism of ordinary psychology provides a basis for distinguishing between cases of genuine intelligence and fakes. And it is one more nail in the coffin of philosophical behaviourism.

You might still be inclined to think that this robot is intelligent in a way. After all, people commonly ascribe intelligence to many robots, including, for example, the car-building ones that are capable of making very fine adjustments to the components of cars, which they go on to assemble. Part of the trouble is that we are bumping up against the vagueness of our ordinary notions. The machine table robot has all the behavioural dispositions of a normal human being, so if that is enough for intelligence, then it is intelligent after all. Even if I am right in thinking it doesn't have the basic package and is not really intelligent, we can hardly avoid treating it as intelligent for practical purposes. The fact is that our everyday conceptions have not been trained to cope with such an abnormal system. Faced with it, ordinary assumptions give out. But that need not bother us. There is surely no reason to assume there is a Platonic Form of intelligence, discoverable if only we work at it hard enough. Intelligence is our own concept, evolved over many centuries (recall Hephaistos' robot servants with *nous*). We should not be surprised that it doesn't enable us to decide all possible cases, no matter how bizarre they may be. Of course we could introduce definitions designed to deal with special cases: perhaps systems like the machine table robot would count as only 'behaviourally intelligent'. But for our purposes that is unnecessary, bearing in mind that our main focus is on consciousness and whatever sort of intelligence that requires. I have argued that it requires the basic package, which the machine table robot doesn't have.

9. The basic package is not enough for consciousness

The basic package is necessary for consciousness and intelligence. But is it sufficient too? Recall the remarkable condition known as 'blindsight'. In some people, half the visual field seems to have been blotted out (typically on account of brain damage): they have no visual images in the affected left or right halves of their eyes. However, some of those who have this condition nevertheless get some limited information about what would have been visible only in their supposedly 'blind' areas. Since they typically claim to see nothing there, they usually have to be persuaded to guess. And – this is the point – they are quite successful in answering forced-choice questions about stimuli which have been arranged to be exposed only to their blind areas. When offered two alternative descriptions of the current stimulus and asked to guess which is correct, it turns out that they perform substantially better than chance. This proves that some information about the stimulus is being received via their eyes in spite of their claims to 'see' nothing in the affected areas. If we are to believe what they tell us (and who are we to contradict them?) it looks as if the basic package is not enough for conscious perceptual experience. Blindsighters acquire information via their eyes about items in their blind fields, yet according to them they have no associated conscious visual experiences.

Fortunately we don't have to decide whether or not to believe blindsighters' reports; a different example will be enough to make the point. Suppose you are asked what you ate for lunch yesterday, or where you were last Tuesday. Possibly you can say straight away; but if you are like me you will need time to think, and may still get it wrong. We usually have to recollect, or summon up, that sort of information; we have received it and retained it, but it doesn't immediately act upon us; it isn't 'forced' on us. And before we can use it to guide our actions – for example to reply to those questions – it has to be called up. The information in question here is just the sort we normally acquire by perception. And we can have information of that kind even when it is not being presented to us as perception normally presents it: consciously.

These thoughts suggest that there might have been creatures which acquired just that sort of information by a *kind* of perception – but not by normal conscious perception. So we can imagine there might have been for example an animal superficially like a rabbit, with a rabbit's normal appearance and appetites, but for which information it acquired via eyes, ears, and other senses was not conscious, but just stored in its brain and available to be used only if the animal called it up, or perhaps if a piece of that information just happened to 'come to mind'. This animal (a mutation from normal rabbits, perhaps) would have all the capacities in the basic package – it would be a decider in my sense, just as I assume normal rabbits are – yet it would not be conscious.

An animal of that kind – a 'rabbitoid' – would obviously be at a huge disadvantage compared with normal rabbits. Suppose a rabbitoid was in a lettuce patch and, in the unconscious manner just explained, acquired the information that it was surrounded by lettuces. And suppose a fox entered that lettuce patch. Information about the fox's arrival would get into the rabbitoid's brain, and would be available for use provided either the animal called it up, or something caused it to pop up. But the point is that nothing would force it to call it up, nor would it necessarily pop up. So very likely this creature would be caught and killed – when a normal rabbit would have spotted the fox and retreated to its burrow. The rabbitoid lacks the behavioural capacities of a normally conscious creature. (This contrast between rabbitoids and normal rabbits incidentally illustrates the evolutionary value of consciousness.)

Although that is obviously important, the main point is that the rabbitoid would still be a decider, with the capacity to acquire perceptual information that was *for* it. If that is right – and I shall assume it is – then while the basic package is necessary for perceptual consciousness, it is not also sufficient. We need to consider what else is needed.

9

What's needed on top of the basic package

1. The thesis

The rabbitoid is a peculiar kind of creature. It looks like a normal rabbit and in many ways behaves like one, yet it lacks something necessary for perceptual consciousness. When normal rabbits are going about their business, incoming perceptual information is immediately available for use – and they cannot prevent it from being so. With rabbitoids, in contrast, the information is just fed into memory and takes its place among current beliefs; from there it might be called up or just pop up at random. I can now state the main thesis of this book: a system is perceptually conscious if and only if it has the basic package and cannot choose to prevent incoming perceptual information from being available for immediate use. In the course of this chapter, I will try to make this thesis reasonably clear.

2. The presence of perceptual information

You may recall something said earlier about information. The acquisition, storage, and retrieval of information consists of *changes* in the system's internal states, changes which affect a whole range of internal processes and contribute to guiding its behaviour in ways related to whatever interactions with the environment caused the changes. (The changes in our case, of course, consist of electrochemical reactions among billions of neurones.) And now it is important to keep in mind that events describable in one vocabulary are often redescribable in others. In particular, descriptions such as 'interpretation', 'decision-making', and 'perceptual information' apply to processes and states also describable in terms of the states and interactions of neurones and neurotransmitters – and their applicability doesn't require anything in addition to those low-level processes; they are just redescriptions. Talking in these terms is a way of talking about what's happening down among the neurones, or in the case of artefacts, among such things as electronic circuits. In particular, talking about acquiring perceptual information is a way of talking about changes occurring at those low levels, in spite of the fact that most of us are ignorant of which low-level processes are involved, just as people talked about lightning and eclipses ages before anyone had investigated electricity or discovered the actual motions of the sun and moon.

I will say that when a system with the basic package – a decider – cannot prevent incoming perceptual information from being available for immediate use, the information is 'present'. (In other works I have used 'directly active' instead of 'present'.) The presence of information in this sense shows up in indefinitely many ways. In particular, it doesn't have to be called up; doesn't have to be waited for until it comes to mind; doesn't have to be guessed at, as it does in the case of blindsight. Those negative features are consequences of the fact that the information is available for immediate use no matter what the system may do, short of blocking its eyes, ears, or whatever. For most of us in normal circumstances, the flow of incoming conscious perceptual information is not only continuous and abundant; it is there for us to use regardless of what we might want. The evolutionary benefits of the presence of incoming perceptual information are obvious. When a creature acquires information in that way, it is among other things well placed to revise its goals as soon as its situation changes; we might say the information prompts it to consider revising its current goals. And because the information cannot be blocked, it works on the system's processes of interpretation, assessment, and decision-making regardless of any relevance it may have to its current goals. The point is not that the information *necessarily* affects current goals, or necessarily causes any kind of action. It is that it instantly affects those central processes of interpretation, assessment, and decision-making in ways which enable the system to modify its goals if it chooses. Unconsciously acquired perceptual information doesn't do that.

So my thesis is that to be perceptually conscious is to have the basic package plus presence. But I am a long way from a complete defence of it.

3. Is this stuff scientifically respectable?

One thing that might make you uneasy is that my descriptions of the basic package and presence are vague. Another is that if these claims are already supported by scientific research, aren't they superfluous? If not supported by research, aren't they just idle speculation?

The vagueness of descriptions of the basic package and presence is of a piece with the general vagueness of everyday psychology. This was developed in communities with virtually no scientific knowledge, for purposes remote not only from those of physicists, chemists, and neuroscientists, but also from those of modern psychologists. Its concepts evolved in order to describe, explain, predict, and understand human behaviour in terms that could be easily grasped by people with no conception of science and no knowledge of scientific ways of verifying or falsifying their opinions. That means its concepts tend to be applicable purely on the basis of our capacities to detect patterns of behaviour, together with the fact that some of these behavioural patterns are associated with certain kinds of experience (I winced because the dentist's probe hurt). That in turn means they can be applied to indefinitely many different kinds of systems other than human beings – even to extraterrestrials, regardless of all the details of their internal workings.

Perhaps our everyday psychological concepts will eventually be replaced by scientific ones that fit the facts better. At present, though, we seem to have no concepts to do the work of the notions I have been using, even though they are inevitably blunt instruments, not precision tools. But since the concept of consciousness itself is also pretty vague, an ideally clear account of it may be too much to hope for. In any case, I claim the approach I am developing provides at least useful suggestions for explaining how there can be such a thing as perceptual consciousness. Vague and blunt though the notions of the basic package and presence are, they may still get us on to the right track. A further important point is that because these notions can be applied to a whole range of systems, regardless of the details of their construction – they might not share our biology, might not even be biological – it doesn't matter if we cannot give detailed neurophysiological details of what actually performs the relevant functions. It would be immensely valuable to know what mechanisms were involved, of course, but so far as I know we don't quite have that knowledge yet, even though this area of research is being intensively developed.

Still, although my account is not itself scientific, it had better be consistent with existing scientific knowledge and also with a scientific approach: no covert appeal to magic. Contrast the epiphenomenalists, who maintain that although qualia don't have effects on the physical world, we can still talk about them – which, as we saw in Chapter 7, would surely have to depend on magic: we could hardly talk about things that didn't affect our behaviour. So how might presence be investigated scientifically?

4. Investigating presence scientifically

We need to discover both the given system's dispositions to overt behaviour and its internal processes. Are there any obstacles to a scientific investigation of these matters? Obviously there will be borderline cases, and hard cases where a system's sensory equipment does not match that of the investigators (as with bats). But what scientific respectability requires is that, given time and ingenuity, the question is answerable – for normal and standard cases at least, if not for all imaginable ones. As typically in scientific research, the investigators will reach a verdict by interacting with the system and observing its behaviour and internal processing in a range of situations.

Think of what happens if you see a meteor. A point of bright light moves rapidly across your retinas, causing sequences of electrochemical signals to go along your optic nerves and into various parts of the brain, including the primary and secondary visual cortex. Further signals are transmitted to complex neural structures that are said to construct representations of what is seen: interpreting and classifying the patterns of neural activity as these evolve through time. When these and related neural processes occur, they are redescribable as your *receiving perceptual information*. But now, in the light of our thoughts about the rabbitoid's non-conscious visual perception, it seems conceivable that you should receive such perceptual information without actually having a conscious experience of the meteor. With that example and those thoughts in mind, my claim is that if the incoming perceptual is in fact conscious, what makes it so is its becoming immediately available for you to use. The neural events constituting your acquisition of that information include events which you cannot prevent from enabling you to act on it: to say, for example, 'Did you see that?'.

My guess is that it is in principle scientifically discoverable whether or not events of the sort alluded to in the last sentence are occurring, in particular events and processes that are redescribable as ensuring that you cannot prevent the information being immediately available for you to use. Evidently I have only half-baked ideas on this subject, and may easily be wrong. But those are my reasons for thinking the line I am taking is scientifically respectable.

5. A dilemma?

Back now to the main thesis. When attacking the 'official' dualistic conception of the mind, Ryle argued that it depends on an unworkable model of perceptual experience. This conceives of a person's mind receiving and interpreting internal (non-physical) images or representations deriving from the senses – like viewing TV. Ryle's objection to this Cartesian Theatre model was devastating. According to the model, a perceptual experience consists of an episode of *internal perception*. When one sees a fox, for example, there is an internal item before the mind (an image or 'sense-datum' of a fox). In order to explain what this involves, we are told that the mind internally perceives the image. Now, this model might be useful if we already knew what internal perception involved. But the model supplies no explanation of that. Of course there are various ways in which it might be amplified. But Ryle's point is that if this account requires internal perception to be importantly like normal perception, then it doesn't explain anything. It just presupposes we already know what perception is, and tells us that ordinary perception involves internal perception. Consider: if internal perception requires a little perceiver inside the head which receives inputs from the senses, then it seems that this homunculus's perception must in consistency involve a further-back little perceiver, and so on ad infinitum – which means there is no point where internal perception actually occurs. If on the other hand internal perception is significantly different from normal perception, then we don't know what it is and again the model fails to explain perception. Either way, the Cartesian Theatre model explains nothing.

Clearly, then, if the notion of the presence of incoming perceptual information is to be any use, it must avoid that model – but you may suspect it comes dangerously close. Don't I require the system's processes of assessment, interpretation, and decision-making to be directed at incoming perceptual information? The seeming dilemma is this. Either the incoming perceptual information is faced by a central system, or it is not. The first alternative appears to depend on the useless Cartesian Theatre model. The second provides for no central system – which seems to imply that the incoming perceptual information has nowhere to go and streams away uselessly into the void. It therefore looks as if neither alternative could be correct, in which case we must reject both. If that is right, I have not begun to explain conscious perceptual experience and cannot even say what a scientific investigation would have to look for.

Fortunately there is no real dilemma. We don't have to choose between those models. There is a third option, which comes into view when we recall that the capacities involved in the basic package are interdependent. As we saw in the previous chapter, there is in general no such thing as a system having information which it can use unless it can do a number of other things. For example, it must be capable of being guided by the information, which means it must have goals and the capacity to store and retrieve information; which in turn means it must be able to initiate and control its own behaviour (see especially

8§5). In general, no internal processes considered in isolation from the rest can be counted as 'assessment', 'interpretation', or whatever. These expressions are redescriptions of sub-processes in a whole complex, and apply only when the system has all or most of these capacities. Now, the rabbitoid helped to show that the capacity to receive perceptual information with presence is not necessary for the basic package. On the other hand, the basic package is necessary for present perceptual information, which means that the presence of perceptual information depends on the interdependent capacities in the basic package. That in turn implies there is a great swirl of processes in which no component sub-process *by itself* constitutes anything like noticing or interpreting or being aware or conscious of incoming perceptual information. Without the whole complex, there is no conscious experience at all. The complex is *redescribable* as including those activities, together with assessment, decision-making, and the other components of the basic package.

I conclude that the apparent dilemma does not threaten the thesis that a system is perceptually conscious if and only if it is a decider whose incoming perceptual information is present.

6. Conclusion

I have been saying that perceptual consciousness involves the reception, storage, and interpretation of incoming information, the subject's assessment of its situation and consequent decision-making, and – crucially – some of that information being in a special sense present. However, although these descriptions mention what is *done*, they conspicuously fail to tell us about what does the doing – for example, which kinds of neurones or regions of the brain are involved. Is that a mistake? No. My account is a functionalist one, and such accounts don't need details of how the relevant processes are realized. However, there is much more to say about functionalism, and plenty of objections to be considered.

10

Functionalism

1. Functionalism and causation

All functionalists agree that having mental states is a matter of the performance of certain functions. It doesn't matter what performs these functions; all that matters is *that* they are performed. In this chapter I shall say more about functionalism in general and explain the distinctive features of my own version.

The most common variety of functionalism rests on the notion of causation. The idea is that each type of mental state is characterized by causal relationships between sensory inputs, behavioural outputs, and other mental states: by what kinds of inputs and other states cause it, and what kinds of behaviour and other states it causes. A first stab at a causal functional definition of pain might be that it is typically caused by damage to the subject's body and typically causes the subject to wince, groan, or scream, depending on the severity of the damage. That is surely too crude as it stands, but it does offer a rough basis for distinguishing pain from other types of mental state, and it illustrates the point of a functionalist approach. For it imposes no conditions on what actually performs the causal functions in question. It doesn't even require them to be performed by brain processes – they might be performed by, for example, electronic devices.

Functionalism contrasts strikingly with the identity theory (5§8). To take a hackneyed and wrong but still useful example, it has been suggested that pain is *identical* with the firing of C-fibres (a kind of neurones). If pain really were identical with C-fibre firing – indeed, if it really were identical with any kind of physical process – then functionalism would be false. By the same token, if the functionalist approach is correct, pain is not identical with any such process. According to functionalism, two systems might both be in pain even if they had no physical features in common. The functions in question might be performed in you by the firing of C-fibres; in me by the firing of D-fibres; in Hephaistos' servants by certain patterns of activation of elements of their electronic brains.

But many people think functionalism is crazy. They have the intuition that a system might be exactly like one of us as far as the functions performed in it were concerned, yet either without sensations altogether or else with different sensations from ours – for there doesn't appear to be any necessary link between sensations and the performance of functions. At any rate there doesn't appear to be a logically necessary link, which is what matters here, given that we are trying to discover *what it takes* for a system to be conscious, not merely what is correlated with consciousness.

We have met something like that objection before, of course: the idea of zombies. The suggestion was that something might be exactly like one of us as far as the *physical* facts about it were concerned, yet either without conscious experiences altogether (a zombie) or else with different experiences (transposed qualia). I objected that the zombie idea depends on mistaken assumptions: notably the assumption that consciousness is a kind of stuff such as paint, or an object like a jacket. However, the notion we are now focusing on is that of a *functional* zombie, which is different from the notion as I have defined it and as it is usually understood; similarly, a case of transposed qualia in two systems that shared the same *functional* organization would be different from transposed qualia in two *physically* similar systems. For it is at least not absurd to suggest that non-functional differences might account for a difference in respect of consciousness between two 'functionally isomorphic' systems: systems exactly similar in functional respects. I will discuss functional zombies and functional transposed qualia in the next chapter, arguing that – this may seem surprising or even shocking – we must all be functionalists. For the present, you may want to suspend judgement.

If we temporarily sidestep the question of how consciousness could possibly be just a matter of the performance of functions in the first place, sensations such as pain seem to fit the causal-functional approach quite well. You might still wonder how it deals with other kinds of mental state such as beliefs and desires. Can there really be a pattern of causes and effects distinctive of my believing that there are rabbits in Wales, for example? True, I am normally disposed to reply 'yes' if asked 'Are there rabbits in Wales?'; to use that statement as a premiss in arguments about the distribution of rabbits, and so on. At first that may seem to show my belief can be accounted for in terms of a relatively limited range of causes and effects. But this is quickly seen to be a mistake: I might behave differently if my interlocutor wanted to shoot rabbits and I wanted to thwart him: there is no characterizing belief in terms of relatively simple patterns of cause and effect. Further difficulties arise when we turn to consider the causes themselves. No doubt definite facts, such as my having actually spotted rabbits in the Wales, or being told about them by a reliable witness, caused me to acquire the belief in question. But there seems no limit to the number of ways in which I might have acquired that same belief, so it seems that having it cannot be the same thing as its having been caused in just that way, with just those behavioural effects. It looks as if functionalists of the purely causal variety cannot maintain that each belief is characterized by a distinctive pattern of causes and effects. They seem forced to retreat to a more general claim on the lines of: 'each mental state is characterized by its causal relations to a complex network of sensory inputs, behavioural outputs, and other mental states'. If causal functionalism is understood along those lines, it starts to seem quite plausible.

2. Functions

The brand of functionalism I am advocating does not necessarily reject the causal functionalists' claim, but doesn't use causation as the basic relation. Instead it uses the notion of a function as something that is *done* by something. (In our own case, the things in question are done by neural processes.) But it is not clear how we should decide what a thing's function is. Unlike size and shape, a thing's functions cannot be read off from its appearance or description. If we divide things into artefacts and others, the functions of artefacts are usually easy to discover: determined by their makers' intentions. A chair is intended to be a portable seat for one; a clock's function is to tell the time, and its hands are there to indicate hours and minutes. Not that we have to use things to do what they were designed for: the chair may serve as firewood; the clock could serve as a missile; a spare minute hand might be used as a spoon. But the link to intentions usually makes it fairly easy to establish an artefact's functions. (Intentions may be eccentric, though. My computer is making a slight whispering sound. Is that one of its functions? It would have been if, bizarrely, I had bought it just for the sake of that noise.) As for natural objects, many have functions definable relative to human interests: this is a guide-dog; this piece of ash is a walking-stick, and so on. Evolution provides a useful basis for assigning functions to features of biological systems, on the assumption that the presence of a given feature in an evolved organism is normally explicable in terms of its survival value. What's the function of the heart? To pump blood around the body. And what's the function of blood circulation? To supply nutrients to organs throughout the body.

Some things that are neither artefacts nor natural objects perform functions even if these are not fixed by – or even in accordance with – anyone's intentions. Social institutions are one example: tabus against sibling marriage reduce the chances of harms resulting from inbreeding even if the population doesn't know it.

Evidently, what we classify as a thing's functions depends largely on our interests. We are interested in making things to serve our purposes, hence those obvious points about the functions of artefacts. We are interested in understanding how naturally evolved systems come into existence and develop, hence our use of evolution as a basis for assigning functions to such systems. We are interested in the dynamics of social systems, hence our assignment of functions to socially established practices.

The functions specially relevant to the brand of functionalism I recommend are characterizable in terms of the basic package and presence: acquiring, storing, and retrieving information; modifying behaviour on the basis of incoming perceptual information; and so on. I take it that the idea that these functions are in fact performed is part of our everyday or folk psychology – which, as we have noted, is something no one has so far managed to do without, and which (some details apart) many think we might never be able to do without. I take it that our interest in the performance of these functions is explicable on the basis of our interest in understanding ourselves and others.

It is worth re-emphasizing that there is no limit to the number of different ways in which systems with the basic package and presence might be constructed. For example, they don't have to be carbon-based; silicon-based systems might do relevantly similar work. In radically different possible universes from ours, even different physical laws would provide for systems that were fundamentally different from terrestrial systems in physical respects; yet they might nevertheless include deciders with presence. Since what I claim are the 'right functions' would still be performed in such systems, the latter would be perceptually conscious, or so I am in the course of arguing. (Let me add that I know of no reason why the relevant functions should not be performed by non-physical systems.)

3. Functionalism must be deep

The Giant illustrates not only the failure of behaviourism, but also of what may be called 'surface' functionalism. The Giant's internal workings consist of the activities of its controllers; so it is like a monstrous puppet and cannot have its own thoughts or feelings. The machine table robot teaches much the same lesson, even though it isn't a puppet because it has the kind of autonomy that the Giant lacks. The trouble with the machine table robot is that nothing going on inside it could contribute to its being aware of things like books or buzzards. This is because at any instant all its sources of sensory information are represented by a single number, so there are no processes that could qualify as its being aware of the book's colour as distinct from the buzzard's cry – when moderate realism requires there to be such processes. It seems, then, that if a functionalist account of consciousness is to have a chance of success, it must be *deep*, in the sense that it imposes certain constraints on the nature of a system's internal processes.

As suggested earlier, I think these constraints are implied by the moderate realism of everyday psychology. They are adequately represented by the requirement that only deciders – systems with the basic package – can have psychological states (bearing in mind that only deciders can have the additional feature of presence). Some objections to functionalism work only against surface varieties.

4. Mental states need not be functionally definable

A common objection to functionalism is that it requires mental states to be *definable* in functional terms, when no such definitions seem to be available. Statements about the acquisition, storage, and retrieval of information, even taken together with statements about interpretation, assessment, and the irresistibility of inputs, just don't seem capable of providing equivalents to statements about conscious experiences, hence don't seem capable of providing definitions of them. When I describe my experience of seeing the waves churning in the sea, expressions such as *grey-green, foaming,* together with the phenomenal concepts we typically use to describe what our experiences are like, come from regions of the conceptual repertoire remote from the home of concepts like *information, assessment*, and the rest. And if I try to restate my description of the experience in those very different terms, I am baffled. The fact is that phenomenal concepts are different in their very nature from those used to characterize the performance of functions.

For phenomenal concepts require their users to have a certain kind of knowledge – knowledge of *what it is like* to have experiences – while concepts specifying functions do not. I cannot say what it is like for me to see a red rose without somehow using the concept *red*, and I cannot have that concept unless I have had certain kinds of experiences, probably including actually seeing red things. Such concepts may be called 'viewpoint-relative' (or 'first-person' concepts; the idea is from Nagel 1986). On the other hand, the concepts required to describe the acquisition, storage, and retrieval of information can be acquired without their users needing to have had any particular kind of experience; they are 'viewpoint-neutral'; and the same is true of a much wider range of concepts, for example those expressed in the vocabulary of physics. This is a solid reason why concepts for characterizing functional roles cannot be used to construct expressions equivalent to ones using viewpoint-relative concepts such as 'looks blue'. No possible viewpoint-neutral (or 'third-person') definitions could characterize such feels. Therefore no functionalist approach could possibly succeed – if functionalism did indeed require definitions based on such characterizations.

But does it? Consider other cases, for example those where things describable in one vocabulary are composed of things describable in radically different vocabularies. Houses are built from, say, bricks, mortar, pieces of wood, and tiles. But houses may be described as spacious or cramped, beautiful or ugly: descriptions without equivalents in terms of building materials. Yet it's not as if the houses consisted of anything over and above bricks, tiles, and the rest, suitably arranged. Descriptions such as 'spacious' and 'beautiful' apply on account of features that the houses have as wholes (on account of certain relationships among their parts). They don't necessarily have to apply to their components. Indeed, we can use whole-house descriptions correctly without having any idea what houses are made of. Two other examples: the locations and chemical composition of the colours used in a painting contrasted with descriptions in pictorial terms; and facts about which words occur in a poem in which order, contrasted with interpretations and other literary comments.

Functionalists can say that something similar holds for the relation between conscious experiences and the performance of functions, although of course this relation is not like that between a house and its components. In spite of the fact that my experience of seeing the foaming grey-green waves is a matter of the performance of certain functions by processes in my brain, it doesn't follow that descriptions of my experience must be translatable or definable in terms of the performance of functions.

Given there is no more to a house than its components suitably arranged, those components and the particular way they are arranged in a particular building will be *sufficient* for the truth of the other descriptions in spite of the very different vocabularies involved. If a building satisfies a given specification of its components and their relations, then it can hardly fail to satisfy those other descriptions too. But we are considering an objection against the possibility of functional *definitions*, and a definition requires conditions that are necessary as well as sufficient. After all, a full description of the entire physical universe, specifying the locations and states of all elementary particles at each instant throughout the whole of time, would be sufficient for the existence of all the houses there have ever been, and also for the truth of whatever other descriptions will have been true of them. But presumably a specification of the entire universe is not also a necessary component of a definition of a house. Similarly, functionalists can maintain that although the performance of certain functions in me is sufficient for my having this experience of the churning grey-green sea, it doesn't follow that it is also necessary for the same experience. We can insist that indefinitely many different combinations of functional states would also have been sufficient. For that reason, as well as the one given earlier, we don't have to accept that the experience must be definable in terms of functional states – even though it doesn't involve anything beyond the performance of functions.

Some people imply that functionalism entails that mental concepts are 'functional concepts'. I am hoping this section blocks the inclination to hold that view. Certainly, we use functional concepts such as *information, memory, storage*, in order to specify the kinds of functions whose performance constitutes conscious experiences. But for the reasons given in the earlier discussions, it doesn't follow that the first-person concepts used to characterize the experiences must also be functional.

5. Two more objections

Another objection arises from reflecting on the 'feel' of experiences: 'You say consciousness is just a matter of states and processes performing certain functions. But as Chalmers emphasizes, the fundamental point is that what makes states conscious "is that they have a certain phenomenal feel, and this feel is not something that can be functionally defined away" (1996: 105). Functional states can't have feels, this objection continues; even if they are associated with feels, they could occur without them, so feels can't be essential to them.'

Reply. Functionalism – at any rate my version – is very far from attempting to define consciousness *away*. It aims to explain what consciousness *is*. Nor, as I have just pointed out, does functionalism require definitions of feels. I will not challenge the assumption that feels are 'essential' properties of conscious states. But I do challenge Chalmers's assumption that conscious states cannot *be* functional states. If my toothache is a functional state, and its feel is essential to it, then that feel is essential to that functional state, counter-intuitive though that may be. (I hope to remove the implausibility: keep reading.) Chalmers's reasoning can be put like this. Conscious experience – 'first-person experience' – is something we know about by actually having experiences at first hand. But the way we learn how our cognitive systems function is different. We start with observation; we learn it 'from the third-person point of view'. So 'what needs to be explained [Chalmers concludes] cannot be analysed as the playing of some functional role, for the latter phenomenon is revealed to us by third-person observation and is much more straightforward' (1996: 110). But that argument overlooks a vital consideration. Although it is more straightforward to understand how functions are performed than to understand how conscious experiences occur, it doesn't follow that there is any more to conscious experience than the performance of functional roles. Chalmers fails to take seriously the possibility that what at first seem like two different phenomena may turn out to be the same. He asserts that 'the problem of consciousness goes beyond any problem about the explanation of structure and function, so a new sort of explanation is needed' (121). He is right to point out that explanations of structure and function will not do the necessary work by themselves; they need to be supplemented by philosophical argument – which is what I am trying to provide. It is also true that something like a new sort of explanation is needed. But he is wrong to assert that such an explanation requires, as he claims, 'radical changes in the way we think about the structure of the world' (122). Nothing he says rules out the possibility of explaining consciousness in functional terms. (Nothing, that is, apart from his fatal commitment to intuitions about zombies and transposed qualia.)

Yet another objection: 'You are ignoring a vital fact. The performance of functions is a fundamentally different kind of thing from the occurrence of conscious experiences. Functional concepts apply on account of relations between different informational concepts (or other kinds of mental states) while the phenomenal concepts that we use to characterize feels apply on account of the intrinsic character of the experiences they refer to. We could say that while functional concepts apply on account of what the various interrelated states *do*, phenomenal concepts apply on account of what the various states of consciousness *are*. You can't characterize what they are by pointing to what they do. That is why phenomenal properties cannot be explained in terms of functional properties.'

Reply: That last sentence gives the game away, it simply begs the question against the view I am advocating. Certainly the concepts are different in nature, and that is what rules out functional *definitions* of conscious states. But it doesn't rule out functional *explanations* of conscious states. The key consideration is this. The functional explanation aims to show how there can be conscious states at all: how there can be such beings as conscious subjects in the first place. It follows that if this explanation is successful, it will at the same time explain how there is scope for suitably intelligent conscious subjects to develop their own special ways of thinking and talking about their conscious states: to have their own points of view. One significant thing to note is that in order to do that, they don't have to understand the processes which ensure that they have those states. Given they have those states in the first place, they can learn what they are like by actually having them, rather than by speculating about what underlies them. The concepts they develop for describing their conscious states from their own point of view will be first-person concepts: viewpoint-relative. They will also be concepts which can be regarded as applying to the intrinsic character of the experiences, since they will not be linked in any obvious way to what underlies them, and indeed will seem independent of it. There will be no obvious connection between a functional explanation of what is involved in having an experience of red, and actually having the experience. Nevertheless, a functional explanation can show how conscious states are a matter of relations among processes which perform various functions, while at the same time having intrinsic characters which can be thought about independently of any knowledge of functional facts.

Briefly, I don't need to go to the dentist to know I have toothache, and I don't need to understand the functional basis of toothache in order to be able to acquire the concept *toothache*. That makes it natural to assume that conscious experience is independent of the relationships involved in the basic package and presence. Yet its being natural doesn't make it true. Conscious experience may still depend on processes standing in those relationships.

6. Redescriptions and analyticity

Functionalism implies that truths about mental states – including states of consciousness – are alternative ways of describing what could also have been described in terms of the performance of functions. In that sense they are redescriptions. The idea of redescription has come up earlier in this book and is of course very familiar. 'Twenty-two people are running around in a field and kicking a round ball' – redescription: 'They are playing football.' Another example: pretend it were possible to specify the locations and composition of the molecules composing a certain piece of stone (practical impossibility is beside the point). Then we might be able to redescribe the same piece of stone as a flint 10 cm long, 6 cm wide, and roughly triangular.

Here I want to introduce a special notion of *pure redescription*. The expression '10 cm long, 6 cm wide, and roughly triangular' is a *pure* redescription relative to the original molecular description because its truth doesn't depend on the flint's history or position relative to other things, or on anything else outside what the molecular specification spells out. On the other hand, we might suppose the flint is in fact a palaeolithic axehead, so that 'palaeolithic' and 'axehead' are further redescriptions of it. But because the truth of these two redescriptions depends on historical facts, notably on the flint's relation to our remote ancestors and their intentions, they are not pure redescriptions. One more example: to specify one of the black and white images produced by my digital camera I need only say which pixels are black, using a set of ordered pairs of the form (x,y) (where the variables stand for the coordinates of black pixels). With this simple pixel language I could specify the whole or part of any image producible by the camera. So, we may have a description on the lines of: 'There are black pixels at: (123, 456), (124, 456) …'. One redescription might be: 'The pixels form the image of a reclining cat.' That would be a pure redescription. Another might be: 'The pixels form the image of our cat Zoë' – which is not pure because the original specification leaves it open whether the image is of a particular cat with particular name and owners.

Now, the concepts in the original specification will typically be different from those in any of the pure redescriptions. To use traditional philosophical jargon, there may well not be any *analytic* statements connecting the two descriptions. But I must say something about this idea of analyticity, which is old, problematic, not very clear, yet still sometimes useful. 'Napoleon was exiled to Elba' is a factual statement; what makes it true? That obviously depends partly on what it means: on what the names 'Napoleon' and 'Elba' pick out, and on what 'exiled' means. But it also depends on facts about the world: on whether the individual picked out by the first name was indeed exiled to the island picked out by the second. The idea of analyticity, in contrast, is that some statements are true solely on account of their meanings, no matter how things may be in the world. 'Exile involves being made to live outside one's own country' is an example of an analytic statement; 'Cats are animals' is another. Clearly, the world might have been different in ways that would have resulted in some actually true factual statements being false: they are only *contingently* true. But if the notion of analyticity is sound, there is no way the world could have been different that would have made an actually true analytic statement false: such statements are necessarily true (logically necessary in the sense explained earlier).

Now, the original specification of the stone contains expressions such as 'molecule of silicon at x,y,z', and there are surely no connections between the meanings of those expressions on the one hand, and '10 cm long, 6 cm wide, and roughly triangular' on the other: no analytic connections between them. Yet the latter is still a pure redescription of that same stone. Statements connecting the molecular description of the stone to its description as being of the given length, width, and shape are not linked analytically; nevertheless it is impossible that the first description should have been true and the second false. Given the nature and locations of the molecules which make up that piece of flint, it could not fail to satisfy those descriptions of its shape and dimensions. So, in spite of the absence of analytic connections from one lot of descriptions to the other – there are no individual analytic statements to the effect that if such and such molecular facts hold, then such and such facts about shapes and dimensions hold too – the whole system of truths about the molecules is still necessarily linked to truths about the shapes and sizes of the stone and its parts; there are *holistic* connections between them.

What matters above all is that the truth of the original description fixes the truth of any of the related pure redescriptions. This ensures that the first *logically entails* the second. The explanation is straightforward. We start from the generally accepted fact that what ensures we can use certain words rather than others to describe things is rules, conventions, or linguistic practices. We use the sound (or mark) represented by 'pig' to pick out that particular kind of animal because it is an established practice in our language to use that sound in that way; that practice is what ensures the word has its own particular meaning. Using the sound 'cat' for that purpose wouldn't have turned pigs into cats; it would only have ensured that the picture of a pig in an infants' first reading book would have had the word 'cat' next to it instead of 'pig' – and so on. So the rules or practices of a language ensure that words and sentences have the meanings they do have rather than others, and they make it possible for there to be bits of language which describe things, states of affairs, or other items. Rules of this kind work in two directions: from world to words, and from words to world. Working from world to words, the rules ensure that a given bit of reality can be described in certain ways (e.g. that the flint has that particular molecular description). Working from words to world, they ensure that a given description (e.g. that molecular description) specifies a certain thing or state of affairs (e.g. the flint).

So one description may specify a thing or state of affairs, and that thing or state of affairs may qualify for a particular pure redescription solely on account of what the original description says about it. And that is why it is absolutely impossible for the first description to be true and any of its related pure redescriptions false – it is what ensures that the latter depend for their truth only on what the original description specifies.

I have taken space to go into this slightly technical notion of pure redescription because if functionalism is true, then *truths about experiences are pure redescriptions of events and processes which constitute the performance of certain functions* (namely, those involved in the basic package and presence). That is how the truth of certain functional descriptions can be logically sufficient for the truth of whichever mental truths it may be.

All that may strike some people as preposterous, but I suspect they are still in the grip of the Cartesian intuitions which make the notion of zombies so appealing. Since I claim to have demolished that notion, I can legitimately ignore those reactions. Plenty of difficulties remain, though, even for those of us who have cast off the intuitions; I will face them, including some further objections, in the remaining chapters.

7. Conclusion

In this chapter I have been trying to clarify some of the implications of the functionalism I propose. Any adequate functionalism must be deep: it must impose constraints on a system's internal processing. The relevant constraints, I suggest, are those of everyday psychology's moderate realism, as represented by the basic package. There have to be more or less distinct real internal states corresponding to some of the broad categories of folk psychology (such as attending to a sound, remembering to a smell). It is important that functionalism does not require mental states to be *definable* in functional terms. But if functionalism is true, mental truths, and in particular phenomenal truths – truths about conscious experiences – are pure redescriptions of events and processes which constitute the performance of various kinds of functions. This does not mean there are analytic statements connecting the former to the latter: the mental pure redescriptions are logically sufficient for those mental truths even if the connections are only holistic.But is functionalism correct? And can we perhaps get on without it?

11

Functionalism is compulsory

1. Possible lines of attack

I have argued that since perception involves getting information about the environment, any conscious perceiver must have the basic package, it must be a decider. But I have also argued that this is not enough: incoming perceptual information must be present in my special sense. When it is, the system cannot stop the incoming information from being immediately available to act upon; it doesn't have to guess at it or call it up or wait for it to pop up. My main thesis is that being a decider with presence is both necessary and sufficient for perceptual consciousness. According to the kind of functionalism I advocate, perceptual consciousness *is* performance of the functions involved in presence.

There are two main ways to attack this claim. One is to maintain that I have hit on the wrong functions, the other is to say that consciousness requires something other than the performance of functions. Attacks of the first kind will not bother me. Although I believe the notion of presence is on the right lines, you might be able to persuade me it needs modification to deal with factors I haven't thought of.

The other line of attack – functionalism is altogether misconceived – is a different matter. If it is correct, then although performance of the right functions may be necessary for perceptual consciousness, it is not also sufficient: it's not enough for the right things to be done; what is needed is for the right things to *exist*; some further condition, in addition to the performance of functions, is needed. If this objection is sound, then exactly the same functions might have been performed in two individuals while their experiences had different characters. One difference might be that their experiences were of different *kinds* (transposed qualia); another would be that while one was conscious, the other was not (it would be a functional zombie). In other words, someone in whom exactly the same functions were performed – who shared my functional profile (alternatively was functionally isomorphic to me) – might have significantly different conscious states from mine in similar situations.

I will argue that such a state of affairs is impossible. Evidently the argument for this claim is crucial. If it works, then, taking account of earlier discussions, we had all better take up functionalism. If it doesn't work, my whole position is fundamentally mistaken.

The argument comes in two stages. The first aims to show that we must be functionalists about just being conscious – being *conscious-in-general* – leaving aside what particular kinds of experiences may be involved. The second stage focuses on the facts about different kinds of experience; it aims to show that we must also be functionalists about the particular qualities of individual conscious states (my having *this* sort of experience when I see a blue book). The significance of this distinction will become clear soon.

The arguments in this chapter are offered in support of controversial claims, which is why I am going to try your patience by expounding them at some length. Even so, I shall not attempt to deal with all the difficulties that could be raised; this is still only an introductory work, and you will no doubt think up your own objections. In any case, I am hoping you will eventually agree that even if you resist the reasoning, the conclusion – that functionalism is compulsory – is not actually mad.

2. Argument for functionalism about consciousness-in-general

The first argument is a 'reductio' (like Searle's Chinese Room argument). We start by assuming a certain statement is true, and the argument shows that if it were true, it would be false, in which case it can't be true. In the present case, the premiss for demolition is that a certain human-like system, *k*, shares my functional profile (the same relevant functions are performed in both of us) but is not conscious: *k* is a functional zombie.

There is one other main premiss: having conscious perceptual experiences has effects on my life. This is absolutely vital but will be challenged by epiphenomenalists. However, in view of the arguments and discussions of epiphenomenalism in Chapter 7, I can legitimately ignore their protests and take that premiss for granted.

Now for the argument. Most of us enjoy some experiences and dislike others, and we all compare and talk about experiences. *k* by definition shares my functional profile, and for that reason has all my behavioural dispositions. Therefore *k behaves* as if like me he compared and talked about his experiences. Here is the decisive step: the *causes* of *k*'s behaviour and the dispositions which underlie them cannot possibly be the same as the causes of mine because he completely lacks a whole dimension of life. Being non-conscious, there is *nothing it is like* for *k* to discuss philosophical problems, drink coffee, eat Stilton cheese, have sex, listen to music – nothing it is like to do anything. So, having the experiences involved in doing these things cannot be a causal factor in *k*'s behaviour. Contrast my own case. For me, what it is like to engage in these and other activities – the character of the experiences involved in them – makes an important causal contribution to my enjoying or disliking or talking about them. The point becomes particularly clear when we consider comparison. The concepts we use when we compare experiences depend on how things strike us (*bluish*, *sweet*, for example), and these comparisons involve conceptualized memories often associated with experiences with the same subjective character as the original ones. In the sharpest possible contrast, my assumed functional but non-conscious isomorph *k* could not compare things on the basis of how they struck him. He could not even have concepts that would enable him to classify things depending on how they struck him. For *nothing* would strike him in that way. This means there would be profound functional differences between *k* and me. The existence of these differences contradicts the original assumption that *k* and I share the same functional profile. That assumption turns out to involve a contradiction and therefore could not be true – which is what the argument set out to prove. (If you are an epiphenomenalist or parallelist even in the teeth of Chapters 5, 6, and 7, you will reject those considerations because they depend on the assumption that consciousness has effects on us. But in that case I think you are wrong to reject my earlier arguments.)

In spite of the fact that *k* would say things like 'This is a different blue from that', he would have no idea at all (not being perceptually conscious) of *why* he said those things. By definition, the processes which resulted in those utterances would not include actually thinking about or comparing experiences because he would have no experiences to compare. So the causes of *k*'s behaviour would be radically different from the causes of the same behaviour in me: a huge difference between us – and functional by any standard, however crude. To summarize:

(1) *k* shares my functional profile but (unlike me) is not perceptually conscious: there is *nothing it is like* for him. (Assumption for reductio.)

(2) What my conscious perceptual experiences are like affects my behaviour. (Chapter 7.)

(3) Although (by (1)) *k* behaves as if he were having perceptual experiences, the fact that he isn't really having them entails that the causes of his behaviour and dispositions are radically different from the causes of mine. What *k's* experiences are like cannot be a causal factor in his behaviour as it is in mine. (Follows from (1) and (2).)

(4) The radical differences between the causes of *k*'s behaviour and the causes of mine are functional on any reasonable assumptions, which means that *k* does not after all share my functional profile (as explained in the text).

(5) Therefore (1) cannot be true. (By (1) and (4).)

The assumption that *k* might share my functional profile yet have no conscious experiences has led to a contradiction. If it is true it is false; therefore it is false. So if the same functions are performed in two individuals, then if either of them is conscious, so is the other, and – if the argument works – *we must all be functionalists about consciousness-in-general.* If the right functions are performed, the system cannot fail to be conscious.

You might wonder whether performance of the right functions was necessary as well as sufficient for consciousness-in-general. But that question was answered earlier, when I argued that perceptual consciousness depends on performance of the functions needed for the basic package and presence. Of course idealism, panpsychism and the various brands of dualism all require something in addition to performance of the right functions. But if the argument in this section is sound, it shows they are wrong: performance of the right functions is enough. It is a significant conclusion – my whole position hinges on it.

3. Transposed qualia?

So much for the first of the main argument's two stages: just being conscious-in-general depends on nothing but the performance of certain functions. But is that also true of the actual character of conscious perceptual experiences? Does what it is like for you and me when we see the same colours (for example) depend entirely on the performance of certain functions? Of course one of us might be colour blind, but that is irrelevant because it would show up in objective tests. I am thinking of the philosophical notion of transposed qualia or the inverted spectrum (2§4). The idea is that differences between our experiences in the same situation might be systematically compensated for so that they would not be discoverable by any behavioural tests. Plenty of philosophers who accept functionalism about consciousness-in-general are impressed by the idea of transposed qualia and reject functionalism about particular states of consciousness. They agree there could not be functional zombies, but nevertheless maintain there could be functional transposed qualia: two people with the same functional profile might differ in the specific character of their conscious perceptual experiences in the same situation.

However, if performance of the right functions is sufficient for an individual to have conscious states in the first place – which means it ensures there is something it is like for that individual – what extra could be needed to ensure that the experience had one specific quality rather than another? What could ensure that my experience of something blue was like *this* rather than the same as what it is actually like for me to see something yellow? It is at least natural to suggest that if performance of certain functions is enough for there to be *something* the experience is like, then other functions, more narrowly specifiable, would be enough to fix *what* it is like. Exponents of the view we are now focusing on reject that reasoning. They maintain that functional facts are not enough to fix the specific qualities of experience, and tend to suggest these qualities are fixed by physical facts. That is why my overall argument needs a second stage – although I think it will be fairly straightforward to explain why those philosophers cannot be right. (Some years ago, when these anti-functionalist ideas were originally being developed, the distinction between surface and deep varieties of functionalism was not usually respected – which made functionalism a much easier target than it actually is.)

4. Argument for functionalism about particular states of consciousness

The second argument depends on the fact that functionalism must be deep. The functions to be taken into account include those involved in informational contact: storing and retrieving memories of experiences, attending to them, thinking about them, and comparing them with other experiences. The first thing to consider is that a conscious experience, for example seeing a blue book, does not have just any old effects on the processes involved in informational contact (see Chapter 7). It has a distinctive impact. The impact of seeing the blue book is normally very different from that of seeing a red book (correspondingly, of course, for hearing and the other sensory modalities: we need not consider them separately). So whatever else may be involved in the experience's having an impact of the sort that matters, *experiences with different specific qualities must have different effects* on the processes involved in informational contact.

Further, if a current experience is remembered, then information about it is stored in such a way that the subject can compare it with different experiences, whether or not these are different experiences of the same kind, or contrasting ones. For example, we are disposed to count green as generally more like blue than like red, and blue as very different from orange; the ways in which our information about different experiences is stored and accessed evidently facilitate those dispositions.

So differences between the specific qualities of experiences result in differences between the ways in which information about them is processed. But – here we reach the argument's conclusion – these are differences in what is done, so they are functional. It doesn't matter how it is done: it doesn't matter how the functions are performed so long as they are performed. If that is right, then functionalism is not only right about consciousness-in-general, it is also right about specific states of consciousness. If two individuals in the same situation are having different experiences, there must be a functional difference between them: without functional differences there can be no differences in the qualities of experiences. That is the hinge on which this book turns. Like other arguments, the one just set out may turn out to be defective, but I still think it is strong. To summarize:

(1) Differences between the specific qualities of experiences result in differences between the ways in which information about them is processed. (If there were no such differences, we would not be able to do such things as compare one experience with another.)

(2) These differences between the impacts of different experiences are functional because they are differences in what is done, when *how* it's done doesn't matter.

(3) Therefore, if two individuals in the same situation are having different experiences, there is a functional difference between them – which means functionalism about specific states of consciousness is true.

I said there were philosophers who maintain that physical differences alone would be sufficient for phenomenal differences between people with the same functional profile; they will resist the above argument. However, I don't know how they could reply to it; perhaps I am blind to the merits of their reasoning, although I don't think they have considered this particular argument. Instead of trying to think up replies on their behalf, I will offer further considerations in favour of my view.

One is that physical differences alone will not generally make a difference to the character of an experience. Consider neural processes. Whatever detailed physical facts may underlie the processing of sensory inputs in human beings, they involve neurones firing. And these firings are all-or-nothing: a neurone either fires or it doesn't. What makes it fire on any given occasion is that *enough* neurotransmitter molecules are discharged across a synapse. So long as enough of them cross the synaptic gap, the exact number doesn't matter. Now, the difference between the actual number of neurotransmitter molecules discharged on a particular occasion and that number plus, say, ten would certainly be a physical difference. But since all that matters is whether or not the neurone does actually fire, a relatively small physical difference will not affect the subject's experience. That is one example of physical difference without phenomenal difference. Another: perception depends on the transmission of signals from one component of the system to another; for example from optic nerves to visual areas; or from stored memories to the neural processes involved in thinking. In any such channel of communication it will be possible for one lot of physical processes to be replaced by different ones giving the same outputs for the same inputs. It is useful here to think of the kind of device known as an 'inverter'. The function of an inverter is this: if it receives an impulse, no impulse is sent downstream; if it receives no impulse, an impulse is sent downstream. But the effect of two inverters in series is nil: the second puts out the same impulses as went into the first. So although a pair of inverters in series would constitute a physical difference, there would be no difference in the transmission of a signal from one part of the brain to another, and so (in all probability, given present knowledge) no difference in the experience provided for by such a signal.

In at least some cases of conscious experience, then, the physical details of just *how* neural signals are transmitted make no difference to the experience. If that is right, physical but functionally neutral differences do not necessarily result in phenomenal differences. Still, exponents of the view we are discussing claim that nevertheless some functionally neutral physical differences do make phenomenal differences. But which ones do they suppose make a difference? I can think of no plausible answer (which again may just be a symptom of blindness to the merits of their arguments). So I continue to rest my case on the argument in the first part of this section.

I have argued that functionalism is true both for consciousness-in-general and for particular states of consciousness. I conclude that we must all be functionalists.

5. Transposed experiences with functional differences

You may still suspect that the Lockean idea that different people might have different colour experiences in the same situation is a threat to functionalism. But functionalism actually leaves room for that possibility, and I will explain why. Functionalism allows that two people with the same functional profile for overt behaviour might indeed have different experiences in the same situation – provided there were *internal*-functional differences between them. Imagine that Amy and Bill share the same functional profile, with the important exception that Bill has inverters immediately behind his retinas and Amy does not. (When a neurone immediately behind Bill's retina is firing, it sends no impulse downstream, and when it is not firing, it sends an impulse downstream.) And suppose for simplicity that Amy and Bill are monochromats. I take it the effect of the difference between them is that their respective experiences are related to each other as the positive and negative of a film. When both are looking at the full moon in a dark sky, for example, the pattern of inputs along Bill's optic nerves downstream of the retinas is like the pattern Amy would receive from a black disc against a brilliant white background, and vice versa. So I take it that Bill's experience is like Amy's experience of a black disc against a bright white background, and vice versa. I further assume that this remains true even though Amy and Bill may have differed in that way since birth.

Bill's inverters make a functional difference between him and Amy. It is not just a physical difference, since any number of physically different items could have performed the same inverting function. But that functional difference makes a difference to their experiences. So this case illustrates one way in which functionalists can allow for the possibility of differences in experience between two individuals whose overt behavioural dispositions are the same: we can appeal to the possibility of such an *internal*-functional difference.

You might object that, given that Amy and Bill share the same functional profile, apart from whatever differences result from Bill's having inverters when Amy doesn't, they would collect the same *information* about the world. It is not as if, when both were looking at a pale green door, Amy learned that the door was pale but Bill learned something different. Both would learn that it was pale (remember they are both monochromats). And after all, functional accounts are often given in terms of information-processing functions. Surely, you might think, functionalists cannot appeal to any factors over and above the processing of information. The story of Amy and Bill shows what is wrong with that objection. There can be a relevant internal functional difference without a difference in the information collected. The internal functional difference envisaged would constitute a difference not in the information acquired or possessed, but in *how* it was acquired and possessed.

6. Computers for neurones?

A different line of objection may have occurred to you. The Giant, in spite of having behavioural dispositions essentially the same as those of a normal human being, would not be conscious. But you may say the main point was that the Giant's *internal* functioning was incompatible with the realism of everyday psychology. Why shouldn't there be other kinds of system whose internal workings were consistent with that realism yet still not conscious? What forces us to accept that the right behaviour plus conformity to moderate psychological realism is *sufficient* for perceptual consciousness? As a basis for replying to these questions, let me introduce the *computers-for-neurones* system.

The idea of such a system starts from the fact that each of the billions of nerve cells in our brains is connected to very many others in such a way that the electrochemical impulses transmitted to it from its neighbours can cause it in turn to fire, and also on occasion cause changes to its dispositions to fire in this way. (Notably, if certain nerves connected to it fire frequently, then it becomes readier to fire when they do, and if the number of their firings is reduced, it becomes less ready to fire when they do. These facts underlie learning.) The idea now is that each of these neurones could be modelled by a tiny computer, programmed to have the same dispositions to fire and change as the original neurone. Then each of a person's billions of neurones could be replaced by a tiny computer of that sort. (Not a practical project but, as usual, we are focusing on what is theoretically possible.) The result would be an electronic system exactly mirroring the functions performed by the basic components of our brains. (I slide over difficulties about reproducing the functions of neurotransmitters and neuroinhibitors.) So what if k, who shares my functional profile, were to have a computers-for-neurones system instead of a brain? All my own processes of conceptualization, information storage and retrieval, and the rest would be mirrored in k's metal and silicon brain, so there is no serious doubt that he would be my functional isomorph not only in respect of his overt behaviour, but also in respect of his internal workings. Yet, you might think, he would not be conscious. If that is right, the main arguments in this chapter don't work.

That is an objection to the conclusion of those arguments, not to the arguments themselves. So far as I can see, my arguments do really show that any system which shared my overt and internal functional profile – no matter how it was constructed – would be perceptually conscious, just as I am. To claim that the computers-for-neurones system would nevertheless not be conscious is to ignore the arguments, not to state an objection. If you have an objection to any of the premises, or the inferences from step to step, I would be very interested to know what they are. But a mere appeal to the intuition that the computers-for-neurones system would not be conscious would just beg the question. You might still be left with the thought that there must be something wrong with the arguments, and perhaps you would be right. I will try to deal with these lingering doubts in the course of the next chapters.

7. Conclusion

In Chapter 9 I argued that the basic package plus the presence of incoming perceptual information is necessary for perceptual consciousness. I believe the arguments in this and the previous chapter show it is also sufficient. If functionalism is compulsory – in which case all that matters is that the right functions are performed – then there can be no perceptual consciousness without presence in my special sense, and being a subject of such consciousness just *is* being a decider with presence. In the two concluding chapters, I will defend this view against likely objections.

12

Is there an explanatory gap?

1. A contrast

It remains easy to imagine that whatever is going on inside my head now might have occurred while my experiences were different: indeed that the same functions might have underlain different experiences. Although imaginability doesn't guarantee genuine possibility, you may suspect there is a serious objection to functionalism lurking here. Surely a good explanation ought not to leave room for equally plausible alternatives to be imagined? Shouldn't it set up clear barriers that would expose such apparent alternatives as impossible fantasies?

Some philosophers argue that any satisfactory explanation must be *transparent* in the sense that it must proceed by readily intelligible steps – and that a transparent explanation of consciousness is impossible. There is, they say, an 'explanatory gap'. They are partly motivated by intuitions about zombies and transposed qualia, but since I have argued that these are radically misconceived, I will ignore that particular consideration. However, there is more to the demand for transparency than that. Even if we resist intuitions about functional zombies, the fact that we can *imagine* the same functions being performed while the subject has different experiences may still seem to block a satisfactory functional explanation. Is functionalism doomed after all?

Think of typical scientific explanations of phenomena originally encountered in ignorance of the underlying scientific facts, phenomena that we describe in familiar terms: for example, combustion, boiling, eclipses of the moon. In these cases, once we have grasped the scientific explanation, it seems impossible even to imagine that the relevant explanatory factors might have been in place while that phenomenon did not occur. These scientific explanations are transparent as a functional explanation of consciousness is not. They seem to leave nothing further to be explained. Consider lunar eclipses. Once we accept that light travels in straight lines and that the earth is spherical, we see clearly that when the earth is lined up between the sun and the moon, the earth's shadow will fall on the moon: a state of affairs which just *is* a lunar eclipse. Surely, if we really understand those initial conditions, we cannot even imagine that when they hold there is no eclipse. The same goes for explanations of other familiar phenomena.

The case of consciousness contrasts strikingly with those scientific cases. The latter take us smoothly from initial conditions (the moon's lying between the sun and the earth) to the conclusion (a lunar eclipse). There is no sudden break in the explanation: each step is perfectly clear. In contrast, there seems to be a very conspicuous break between the performance of functions on the one hand and conscious experiences on the other. This is the so-called 'explanatory gap'. According to Joseph Levine, who introduced the expression, 'the principal manifestation of the explanatory gap' is the conceivability of zombies (2001: 79). (Conceivability here means roughly that not only can we imagine the zombies, but also that we can do it in such a way that no contradiction shows up.) He and many others think we can conceive of states of affairs where the supposed physical or functional facts are as they actually are, but there are either no states of consciousness at all or different ones. He says, 'the robustness of the absent and inverted qualia intuitions is testimony to' the functionalists' and physicalists' inability to provide the necessary explanations (1993: 129). Now, even when we ignore the appeal to Cartesian intuitions, the mere fact that we can imagine different experiences associated with the same functions still suggests a genuine difficulty for functionalism. It is that the physical or functional facts alone would not enable anyone, even a being with superhuman intellectual powers, to discover what it would be like to see colours. We touched on this thought when discussing Chalmers's appeal to Jackson's Mary (6§8); let us consider it more closely.

2. From functional to phenomenal

There is a useful traditional distinction between two ways of getting to a conclusion from a set of assumptions. Suppose Holmes knows that:

(1) *Either* the murderer got into the house through the door *or* the murderer got into the house through the window,

and

(2) The murderer did not get into the house through the door.

Then, Holmes can use elementary logic to infer that:

(3) The murderer got into the house through the window.

This little piece of deduction is *a priori* in the sense that the conclusion (3) can be inferred from the premises (1) and (2) without drawing on any information but what those premises themselves provide. Quite generally, an inference from a given set of premises is *a priori* just in case the conclusion follows without appeal to any considerations other than logical. If Holmes's inference had depended on noticing blood on the window-ledge, for example, his reasoning from (1) and (2) would not have been *a priori*. On the other hand, if non-logical facts other than those included in the premises have to be invoked, then the inference is *a posteriori* or empirical.

Using that terminology we can say that the philosophers who think the so-called explanatory gap is a serious problem for physicalism and functionalism rely on the following assumption:

(A) It is impossible to get *a priori* from the physical or functional facts to the facts about what it is like to have this or that experience.

There are good reasons for this important assumption. As we have already noticed, the 'viewpoint-relative' concepts typically used to characterize experiences require their users to draw on certain special kinds of knowledge. In order to have a colour concept such as *blue*, I need to know what it is like to see blue things, which requires me to have had certain kinds of experiences, such as actually seeing blue things (though perhaps an exceptionally acrobatic imagination would be enough). You may recall the case of Mary, brought up in a colourless, black-and-white environment. Jackson argues that she does not yet *know* what it is like to see red, green, or other colours – she lacks the colour concepts – and further that she cannot *acquire* this knowledge *a priori* from her knowledge of the physical facts about colour vision. In contrast, 'viewpoint-neutral' concepts such as those of physics, together with the functional concepts used in describing the acquisition, storage, and retrieval of information, can be acquired and used without needing to have had any particular kind of experience. Mary knows all the physical and functional facts about colour vision; that knowledge is viewpoint-neutral. But she cannot move *a priori* from it to a knowledge of what it is like to see blue things because her viewpoint-neutral scientific knowledge cannot provide her with the necessary experiences.

Not that Mary is completely ignorant of colours. Like blind people, she knows that colours are properties of objects that are easily detectable in normal lighting conditions; that everything visible has a colour; that the colours can be ordered so as to form a solid whose dimensions are hue, lightness, and intensity; that some colours strike us as more or less similar in each of those respects; that ripe tomatoes are typically red, unripe ones green; and so on. She also appreciates the acuteness of Locke's 'studious blind man', whose brilliant suggestion was that the colour scarlet was 'like the sound of a trumpet' (Locke 1689/1975: III.iv.11). But it remains true that she cannot learn *what it is like* to see colours, and cannot have a proper grasp of colour concepts, or a proper understanding of statements about colours, while she remains in her black-and-white environment.

The problem for me now is that, like many others, I think the above reasoning is sound! I therefore have to accept assumption (A). (I do not, however, also go along with the objection to physicalism which Jackson bases on that reasoning.) Because physical and functional concepts are indeed 'viewpoint-neutral' and graspable without having had any particular kinds of experience, knowing physical and functional facts does not necessarily enable one properly to understand or know the truth of statements involving viewpoint-relative concepts such as blue. Since that example is arbitrary, we can conclude generally that the physical and functional facts do not entail phenomenal truths *a priori*. They would have entailed them *a priori* only if knowing them enabled someone to infer the phenomenal truths from them alone. (It is beside the point that we may lack the necessary memory and reasoning powers: if the entailment is not *a priori*, not even a being with superhuman cognitive powers could discover it.)

So there is certainly an important gap, indeed an abyss, between physical and functional facts on the one hand, and the facts of consciousness on the other. We, or at least I, face a weighty question. Is this abyss a deathtrap for functionalism?

3. Separating *a priori* and logical claims

The abyss would have been lethal if it had been logical. It would have implied that the connection from physical and functional facts to the facts of consciousness does not hold by logical necessity – when I have argued that it does. But there is a significant difference between *logical* entailment and *a priori* entailment. If A entails B *a priori*, then A must also entail B logically. But the converse is not necessarily true. A may entail B logically without also entailing it *a priori*.

This is because there is no guarantee that every logically true statement of the form 'If A, then B' can be *known* to be true. Many can, but there are examples which suggest it is not true generally (e.g. mathematical ones such as Goldbach's conjecture that every even number is a sum of two primes). In any case, a statement's just being true is different from its being *knowably* true. If it can be known, that is something on top of its truth.

So, an important step towards showing there is no problematic explanatory gap is to distinguish two distinct claims: (a) that functional truths entail phenomenal truths logically, and (b) that they also entail them *a priori*. A good deal of confusion has resulted from not giving due weight to this vital distinction. I agree with (a) but reject (b). Truths about the character of our experiences are logically entailed by functional truths, but not also *a priori* entailed by them.

The view that phenomenal truths are logically entailed by functional truths is supported by arguments for the view that there is *no more* to having conscious perceptual experiences, each with its own specific character, than the presence of incoming perceptual information (arguments such as those in the previous chapter). I have argued that we must all be functionalists, from which it follows that talking about experiences is an alternative way of talking about the performance of functions; descriptions of experience are *pure redescriptions* of the performance of functions even if no one realizes this. Now, we saw earlier that cases of pure redescription guarantee (broadly) logical connections. If 'There is a mountain at x,y,z,t' is a pure redescription of a certain very large conglomeration of molecules, then a description of that conglomeration would logically imply the description 'There is a mountain'. Therefore there is no possibility whatever that a particular true description of the character of my experience should have failed to apply when this particular lot of functions was being performed. And that means it is by logical necessity that if those particular functions are being performed, then the system is having that particular experience.

So functional truths entail phenomenal truths logically in spite of the fact that they do not generally entail them *a priori*.

4. The explanation: Two crucial facts

Hold on, you may be thinking: the objection is after all that there is an *explanatory* gap between functional and phenomenal facts. Even if the functional facts logically entail the phenomenal facts, we still need an explanation of why they entail just these phenomenal facts rather than others: why they entail that I am having just *this* experience of blue rather than a different one – which is surely imaginable. So we have not managed to get across the abyss yet.

I shall argue that what has been said so far provides the materials for dealing with this worry, but first we need to consider two crucial facts that were noticed but not focused on. One is that we have a kind of access to our conscious states that others cannot have. We know about them by actually being in them, while others can learn about them only by observing our behaviour (or, if neuroscientists, by studying what is happening in our brains – provided, of course, that they know what to look for: something for which I hope this book may be useful). In order to know whether I have toothache, for example, I don't have to consider my own behaviour. I just attend to the experience. The dentist, on the other hand, can tell only by seeing how I react to probing. The dentist can't see or otherwise perceive my pain. No one else has my kind of access to it.

The second crucial fact is this. Not only do we have that special kind of access to our conscious states; in addition, *what* we can know about them in this way – the actual content of our knowledge – is different in kind from what we or others can know about those states from observation. By actually having toothaches, by seeing the clear blue sky, by smelling roasting coffee, we come to know what these experiences are like. If there are also other ways of coming to know what they are like (e.g. via descriptions), they depend ultimately on someone's coming to know by personal experience. And knowing what the experiences are like gives us a very different kind of knowledge from knowing what our behaviour is like or knowing what is happening down among the neurones. Others cannot acquire that kind of knowledge of our own states except by being told.

Those two considerations – special access and special knowledge – explain why phenomenal concepts are viewpoint-relative, hence why phenomenal truths cannot generally be inferred *a priori* from physical or functional truths. And that means they explain why a functional account of consciousness could not be transparent in the sense that other scientific explanations are.

Those considerations are nevertheless consistent with a functional explanation of consciousness. The key point is that in explaining consciousness, the functional account explains the existence of subjects who have a point of view. It explains how there can be beings who, when their sense receptors are stimulated by patterns of light, or sound-waves, or certain kinds of molecules floating in the air, or other kinds of inputs, store and process this information in ways which ensure that there is something it is like for them. What makes it so difficult to accept that such a thing can be explained functionally are certain very natural assumptions, which we find it hard to avoid. We feel forced to think of our experience of the world as like being faced by a cinema show, a succession of images and sounds seen and heard. Thinking of consciousness in that way makes it too easy to imagine that the show might be stopped while all the rest of the functions performed in our bodies continued as before. It encourages a conception of conscious experiences as something like paint, which might have been different, or even totally absent, without any consequences for the rest of the system: the zombie and transposed qualia intuitions. But as my argument against those intuitions shows, this is a misconception.

It fails to take account of the fact that conscious perception cannot involve the presence of a special kind of stuff which might, logically, have been absent. To recall, it involves among other things the storage and retrieval of perceptual information, together with the complicated processes required for being able to notice and attend to present information as it comes in. And on the whole, it is impossible for these processes to qualify for descriptions like 'storing' or 'attending to' information unless they are interdependent as suggested in Chapter 8.

Looking at things from a different angle, a functional account fits in well with the claim that consciousness has evolved. The troubling old question was: How could creatures with this special 'consciousness stuff' have evolved from creatures without it? We replace that with the question of how creatures in which these special kinds of functions were performed could have evolved from creatures where they were not performed. This question is much less difficult because it is usual to describe certain processes occurring in organisms as having evolved to perform various functions, and this is as true of the processing of information as it is of things like digestion and reproduction. So if consciousness is indeed a matter of certain functions being performed, there is no great problem over how creatures with consciousness could have evolved from creatures without it.

You might object that the light of consciousness is either on or off: it is all or nothing. If that is right, it is still a puzzle how conscious creatures could have evolved from non-conscious ones. But I don't think there is a serious difficulty here. For one thing, among those creatures we tend to think of as having some sort of consciousness, it is in many cases limited and impoverished compared with ours. That implies it is a matter of degree (recall Locke's suggestion that even in molluscs 'there is some small dull Perception'). Another consideration is that there are different levels of consciousness between sleep and waking. Finally, there is evidence from research into blindsight, which suggests that consciousness is not all or nothing:

> DB was questioned repeatedly about his vision in his left half-field. Most commonly he said that he saw nothing at all. If pressed, he might say in some tests, but by no means all, that he perhaps had a 'feeling' that a stimulus was approaching or receding, or was 'smooth'

(the O [one of the two visible stimuli presented to this subject]) or 'jagged' (the X). But always he stressed that he saw nothing in the sense of 'seeing', that typically he was guessing, and was at a loss for words to describe any conscious perception. (Weiskrantz 1986: 31)

It seems reasonable to conclude that it is simply indeterminate whether DB has consciousness in his blind half-field.

By showing how there can be conscious states at all – how there can be such a thing as a system with its own point of view – our functionalist account also explains how there is scope for suitably intelligent conscious subjects to develop their own special ways of thinking and talking about their conscious states. To do that they don't have to understand the processes which ensure they have them. Given that they have those states in the first place, they can learn what they are like by having them, rather than by speculating about what underlies them. The concepts they develop for describing their conscious states will be viewpoint-relative, and will also naturally be regarded as applying to the intrinsic character of the experiences, since they will not be linked in any obvious way to what underlies them; indeed they will seem independent of the underlying physical and functional processes. (Which is why Kripke can say that pain is 'picked out by the property of being pain itself, by its immediate phenomenological quality' (1972: 152).) There will be no obvious connection between a functional explanation of what is involved in having an experience of red and actually having the experience. Nevertheless a functional explanation can show how conscious states are constituted by relations among processes performing various functions, even though at the same time they have characters which can be thought about independent of knowing any functional facts. So it will show both why we have that special kind of access to them and why knowledge of experience has its own special character. In that way functionalism (of the right sort) rules out the alleged explanatory gap.

5. Rival accounts

Earlier we discussed some theories which might appear to be alternatives to functionalism. I suggested reasons for rejecting philosophical behaviourism, at least when it purports to explain consciousness; also for rejecting the psycho-physical identity theory. And by arguing that everyone should endorse functionalism, I have incidentally explained why physicalism doesn't need the identity theory, and why panpsychism and various forms of dualism do nothing to explain consciousness. I will say no more about those theories, but I will briefly characterize the main non-functionalist ones, then some alternative broadly functionalist approaches. Although I don't think any of them work as accounts of consciousness, most have something to contribute to our understanding of other aspects of the mental.

Wittgenstein

> But can't I imagine that the people around me are automata, lack consciousness, even though they behave in the same way as usual? – If I imagine it now – alone in my room – I see people with fixed looks (as in a trance) going about their business, – the idea is perhaps a little uncanny. But just try to keep hold of this idea in the midst of your ordinary intercourse with others, in the street, say! Say to yourself, for example: 'The children over there are mere automata; all their liveliness is mere automatism.' And you will either find these words becoming quite meaningless; or you will produce in yourself some kind of uncanny feeling, or something of the sort.
>
> Seeing a living human being as an automaton is analogous to seeing one figure as a limiting case or variant of another: the cross-pieces of a window as a swastika, for example. (Wittgenstein 1953: sec. 420)

Wittgenstein is surely right about the difficulty of imagining the situation he describes, but the point he is making here goes no way towards explaining consciousness. It is not as if exponents of the zombie idea believed we might actually encounter such creatures. All they need is the logical possibility of zombies, which Wittgenstein's remarks do nothing to refute. Besides, he seems to regard attempts to explain consciousness as superfluous or misguided – contrary to what I think has been shown in this book.

Sartre

Sartre says a lot about consciousness in *Being and Nothingness*. He points out that when we look at or otherwise interact with other people, we see them *as* conscious beings (as Wittgenstein also emphasized). We see each person as a 'for-itself' (*pour-soi);* but it is also possible to see them as things: each as an 'in-itself' (*en-soi).* So our attitude to another person is different from, and not reducible to, the attitude we have to a body as an in-itself. He emphasizes that there is no Cartesian mind somehow linked to the body. There is just the body with these two 'aspects': for-itself and in-itself: 'There are no "psychic phenomena" there to be united with the body' (1958: 305). But notice that he just takes for granted that there *are* these special entities, these 'for-itselfs'. He rejects behaviourism, so what does he have to say about the difference between a for-itself and an in-itself? From our point of view it is disappointing. He writes as if there is no problem: no difficulty over explaining what it is for something to be a for-itself. So it seems that he (again like Wittgenstein) just wasn't concerned with the problems that interest us here.

'A sense of self'

Some authors' accounts of consciousness emphasize the role of a creature's conception of itself. Antonio Damasio, for example, says that the simplest kind of consciousness, 'core consciousness', 'provides the organism with a sense of self', and that consciousness in general 'consists of constructing knowledge about two facts: that the organism is involved in relating to some object, and that the object in the relation causes a change in the organism' (1999: 20). These conditions require conscious subjects to be a lot more sophisticated than my account does. Notably, the subjects of Damasio's lowest level of 'consciousness' require the conceptual means to *know that* something outside causes a change in them. On my account that is asking too much. It is enough if the outside world does in fact have relevant effects on the organism; the organism itself need not know what these are. An example is the effects a mouse has on our cat Zoë: she doesn't need to know *that* the mouse has those effects. Nor do I see why all conscious subjects should have a 'sense of self' at all (unless that means only that they must be able to discriminate between themselves and other things – which is true, but doesn't do much to explain consciousness).

The 'extended mind'

Recall that according to externalism, meanings and the contents of mental states 'ain't in the head'; they are determined partly if not wholly by our relations to things in the external world (2§7). The idea is that 'potato' picks out a certain kind of vegetable not because of a special relation to any internal mental notion we may have of potatoes, but because the English-speaking population uses that word to buy potatoes in shops, get other people to pass them at table, and so on. The usual variety of externalism stops at that point. An interesting (though highly controversial) extension of it goes much further. It suggests that our mental processes themselves – our thoughts and feelings – are not in the head either; they are said to be distributed between us and the environment: smeared over both. This is the idea of the 'extended mind'. You might at first guess that this notion is threatened by counter-examples to behaviourism. But it isn't, because unlike behaviourism it doesn't imply that the individual's contribution is limited to behavioural dispositions. It acknowledges that the nature of the internal processing matters too, so that arguments which refute behaviourism will not necessarily also refute the notion of the extended mind. A reason to take note of this approach is that some of its exponents seem to suggest that perceptual consciousness consists *solely* of the organism's ability to use its sensory and motor skills to get around in the world. Now, we must surely agree that perceptual consciousness does require these abilities. They are included in the basic package plus presence, and are at least necessary. But can non-perceptual kinds of consciousness also depend on the same abilities? I shall return to this topic in the next and final chapter (13§§5-6).

Other varieties of functionalism

Throughout I have been taking functionalism to be the idea that mental states consist of the performance of certain functions dependent on relations among a system's inputs, outputs, and internal states, including relations that don't show up in behaviour because they involve only the system's internal processing. Judged by that criterion, the remaining principal accounts of consciousness (to be sketched below) are varieties of functionalism.

Representationalism

If I dream I am driving, my experience *represents* me as doing so. If I believe there are tigers in India, I represent the world (whether truly or falsely) as being a certain way. Nothing specially problematic there. But some philosophers go much further, suggesting that a state's being conscious is *nothing but* its being representational (Dretske 1995; Tye 1995). This is *representationalism* or *intentionalism*. Can it be right?

It faces serious difficulties. Why should the mere fact of a state's being representational make it conscious? Many representational states are not conscious, or not always; my belief that there are tigers in India is one example: most of the time I am completely unaware of it; in fact I think it becomes conscious only on those rare occasions when I use it as a philosophical example. So the claim that consciousness, or having conscious states, is just a matter of these states being representational is implausible. (Of course the version of representationalism sketched above can be varied, but it is hard to see how the necessary further conditions could fail to be functional ones.) Given that I have set out an alternative account which explains consciousness in other ways, I see no need to go along with representationalism. Recall too the example of the rabbitoid. Assuming the notion of such a creature involves no contradiction, as I suggest we can legitimately assume, this creature has lots of representational states – but none of them is conscious. Much more could be said about this approach, but I leave it there.

Inner perception

John Locke had an ingenious theory. He said consciousness was a sort of *inner* perception: 'Consciousness is the perception of what passes in a Man's own Mind' (*Essay* II.i.19). This suggestion was taken up by many philosophers, including Kant, and formed part of the 'official doctrine' criticized so powerfully by Ryle (9§5). The idea is that conscious perception requires an internal perceiver receiving inputs from the senses. The core of Ryle's criticism, you may recall, is a dilemma. If on the one hand inner perception is supposed to be like ordinary perception, then in consistency the internal perceiver's own perception must involve a further-back perceiver, and so on ad infinitum – which seems impossible. If on the other hand internal perception is significantly unlike normal perception, then we haven't been told what it is, and the model again fails to explain perception. Either way, the Cartesian Theatre model is doomed.

However, David Armstrong (1968) argues that consciousness, or rather what he calls 'introspection', is indeed a kind of 'inner sense' and comparable to a computer's scanning its own data. Computers do after all scan data and acquire information about their own internal states, so does this model provide a solid basis for the inner perception theory of consciousness after all? Here we collide with another dilemma. Either the scanned information is fed into some subset of the whole perceiving system, or it isn't. But if all that is going on is the filtering and transmission of incoming information to a subsystem, we have been given no reason why this process should be conscious rather than unconscious (indeed it seems to be what happens in very simple reflex systems, let alone the rabbitoid). If on the other hand the scanned information is not just fed into a subsystem, what is it that actually acquires the scanned information? It must be the system as a whole. The trouble now is that we know that organisms (including ourselves) can acquire information unconsciously, so this horn of the dilemma again prevents the inner perception model from working. What, then, is supposed to guarantee that the system as a whole acquires the scanned information consciously rather than not? No plausible answer is on offer. Either way, this version of the 'inner perception' model breaks down.

Higher-order thought

Locke's broad idea was that consciousness is a matter of the mind somehow apprehending (some of) its own processes; in his account it involves a kind of perception. An alternative development of the same broad idea is to say it involves thoughts about thoughts. (It is assumed that thoughts occur at different levels, so a thought about another thought is a 'higher-order' thought.) The most straightforward approach is David Rosenthal's:

> Conscious states are simply mental states we are conscious of being in. And, in general, our being conscious of something is just a matter of our having a thought of some sort about it. Accordingly, it is natural to identify a mental state's being conscious with one's having a roughly contemporaneous thought that one is in that mental state. (1986: 335)

Take for example my present thought that there is a yellow dog outside. This mental state of mine is said to be conscious just in case I am also having the thought *that* I have the thought that there is a yellow dog outside. There are numerous variations on that simple theme, but all seem vulnerable to three objections. First, it is implausible that *every* conscious thought is accompanied by a thought to the effect that the subject is having it. Second, to have a thought that one is in such and such a mental state demands more cognitive sophistication than babies or languageless animals seem to have, so this theory's exponents are forced to say these creatures are not conscious, contrary to what most of us assume (some of them bite the bullet and accept that counter-intuitive conclusion). Third, no compelling reason has been given for supposing that if a thought is accompanied by a thought about it, then the original thought must necessarily be conscious. Why shouldn't both be unconscious?

Of course, higher-order theorists have replies to such objections, but it would be out of place in this introductory book to pursue what has become a very convoluted debate – especially when I am offering what I believe to be a better approach.

6. Conclusion

I see no good reason to reject my variety of functionalism in favour of one of its rivals. We can avoid falling into what seemed like an abyss between functional and phenomenal facts, provided we can explain why the functional facts logically entail that I am having *this* particular kind of experience rather than some imaginable other kind, and in the previous chapter I anticipated and dealt with this worry. There I argued that if the right functions are performed, then the system cannot fail to be conscious-in-general, and that differences between different kinds of conscious states depend on functional differences, so that in spite of what may be imaginable, the functional facts still absolutely fix what it is like. While there is indeed a gap between functional and phenomenal facts – in that the latter cannot be inferred from the former *a priori* – it is not an *explanatory* gap. The explanation of the failure of *a priori* entailment is entirely consistent with a functional account of consciousness: of both consciousness-in-general and specific states of consciousness.

As we saw in Chapter 6, Chalmers argues that Jackson's Mary cannot learn what it is like to see red solely on the basis of her comprehensive knowledge of the physical facts. He asserts, 'No amount of reasoning from the physical facts alone will give her this knowledge', and concludes that 'it follows that the facts about the subjective experience of colour vision are not entailed by the physical facts' (1996: 103). Since it seems clear that both he and Mary (together of course with physicalists like me) think the functional facts follow from the purely physical facts, that line of argument implies that the facts about subjective experience do not in their turn follow from the functional facts. I hope this chapter shows what is wrong with Chalmers's reasoning. He assumes that if phenomenal truths cannot be *known* on the basis of knowing physical or functional facts, then they do not *follow from* physical and functional truths. I have explained why that assumption, in spite of being widely shared, is mistaken. The functional facts entail the phenomenal facts logically, but not *a priori*.

13

Brains in vats and buckets

1. Where we are

I started to make the case for my kind of functionalism with an examination of Cartesian intuitions about zombies and inverted qualia. These threatened to get us stuck in a bog of confusion, but I think they have been decisively refuted (Chapter 7). Following hints from Locke we moved on to investigate perceptual consciousness, starting with simple organisms. We noted that very primitive stimulus-response systems, though capable of behaviour well able to sustain existence, do not engage in genine learning and therefore do not enjoy genuine conscious experience: they don't even have 'some small dull Perception', as Locke said about oysters. (Was he right? I have no idea, but many creatures have turned out to be more interesting than we had guessed.) So what *does* have genuine conscious experience?

The answer, I argued, comes in two stages. Genuine perception requires genuine learning, and if a system is to be able to learn, it must have the basic package of capacities; it must be a decider. What marks deciders off from other systems is that they can acquire, store, and retrieve information; initiate and modify their behaviour on the basis of incoming perceptual information; interpret information; assess their situations; choose between alternative courses of action; have goals. And those are all redescriptions of (in our own cases) physical processes going on in the system: a swirl of electrochemical events in our brains. They point to the functions that must be performed in a perceptually conscious system. So much for the first stage of my account.

Although the basic package is necessary for perceptual consciousness, it is not sufficient. To introduce the second stage I used the example of the rabbitoid, arguing that such a creature would have the basic package without being perceptually conscious; it would lack a vital feature: the presence of incoming perceptual information. In order to be able to use this information it would have to call it up, or guess at it, or leave it to pop up at random. In sharp contrast, normal rabbits and the rest of us don't have to do any of those things; incoming perceptual information forces itself on us. We cannot stop it being available for our use.

Those reflections suggested the main features of my functional account of perceptual consciousness. I maintain that any decider whose incoming perceptual information is present in the sense explained is perceptually conscious. If that account is on the right lines, it is all we need to explain how perceptual consciousness involves nothing but the performance of functions, even though this suggestion can appear absurd to people who have not worked through the reasoning.

In addition to sketching that version of functionalism I have offered a general defence of functionalism. I argued that two individuals could not have different kinds of experiences unless the underlying functions were different too: same functions, same experiences (Chapter 11). The conclusion – in the teeth of many people's convictions – was that we must all be functionalists: both about consciousness-in-general and about particular states of consciousness.

That the necessary functions are being performed at all is discoverable by observation: the properties involved are 'third-person' properties. But conscious beings like us also have 'first-person' properties, whose character is discoverable by their subject rather than by observers. If there is *something it is like*, then the subject has first-person properties such as that of seeing blue in a particular way. And (it seems) only the subject can discover just what it is like to have those properties. That raises a puzzling question: how is it possible that the very special first-person properties should nevertheless involve nothing but the ordinary and seemingly very different third-person properties? That question may prompt the suspicion that functionalism just begs the question; at least it reveals the need for more explanation.

Although we can use phenomenal concepts in ignorance of the underlying functional facts, those concepts can still *apply* to those functional facts, which can still be what makes the phenomenal truths true. When I look at this blue book, I have a certain particular kind of experience; so, suppose I have a special word, 'kblue', for it. Then that word applies to the performance of whichever functions it may be that constitute that kind of experience – even though I don't know exactly what those functions are, or even that such functions are being performed at all. Yet there is no mystery about why I can use that word to pick out that kind of experience. It is because I am the subject: the one who actually has the experience.

It certainly *seems* as if my concept of that kind of colour experience has no connection with the performance of those functions. On the other hand, the main argument of the book implies that the connection is as tight as it could possibly be: logical in the broad sense explained earlier (6§3).

That it is a logical connection also follows from the reasoning which led to the claim that we must all be functionalists, and in particular to the conclusion that two individuals could not have different kinds of experiences unless the functions performed were different too. That conclusion also helps to remove the threat of the 'explanatory gap'. There is indeed a gap between knowing functional facts and knowing what the experiences are like, but it is only epistemic, not logical. It concerns what is *knowable*, and is explained by the functional account (see in particular 12§4). If there had also been a logical gap, it would have demolished physicalism and functionalism, but (if I am right!) there isn't one.

I don't claim to have solved the philosophical problems of consciousness, only to have given you some idea of what the main problems are and to have sketched one approach to solving them – though I don't know of any serious objections to this approach. In the rest of this chapter I will deal with doubts and queries that may continue to nag, and go on to discuss a question that still leaves me dithering. How should functionalism deal with *non-perceptual* consciousness?

2. Queries and replies

Experiences still seem to be different kinds of thing from the performance of functions.

Reply: Not a serious problem. The moon is a great deal smaller than the sun, but doesn't seem to be. Lightning is the discharge of static electricity, but that is neither obvious nor even graspable except on the basis of a good deal of scientific knowledge. Of course, the case of conscious experiences and functions is not much like the case of lightning and static electricity, but the general point applies in both cases: grasping the functional explanation of consciousness, like grasping the scientific account of lightning, requires a fair amount of work. It is a striking and confusing fact that the *concepts* in terms of which we specify experiences – phenomenal concepts – are unlike those in terms of which we specify functions. But the concepts involved in ordinary scientific explanations are also remote from the concepts in terms of which we talk and think in everyday life. Lots of things have a hidden nature that most of us know little or nothing about, but which has been or probably will eventually be discovered by science. Although the human race has been thinking and talking about fire, air, water, and eclipses of the sun and moon for thousands of years, it is only over the last few centuries that we have acquired a scientific understanding of these phenomena. Familiarity with everyday concepts does not bring even a superficial acquaintance with the scientific concepts introduced to explain the phenomena. You can know in a way what fire is without having a clue about its inner nature. Similarly, you can know in a way what consciousness is without knowing what it takes for something to be conscious.

All the same, it is hard to avoid thinking that not only do phenomenal and functional concepts come from different boxes, but that they relate to different realities.

Reply: Recall two things: the phenomenal concepts we use to talk about experiences were developed in ignorance of the underlying neural processes; and what matters for the use of those concepts is what it is like for the subject. This means that no matter what the physical or functional facts may be, they are irrelevant when we are trying to say what an experience is like. Yes, phenomenal and functional concepts come from different boxes, which is why there are no neat 'analytic' connections from functional truths to phenomenal truths, and why it is impossible to infer phenomenal truths *a priori* from functional truths. But although that makes it natural to guess that the underlying realities are also radically different, I have argued that there are not two realities here at all. We have two different ways of talking and thinking about the same bits of reality.

There is a contradiction at the heart of your position. You claim the phenomenal facts are fixed by functions. But suppose you are now having the particular kind of experience you mentioned earlier, kblue, caused by the blue cover of a certain book. That experience performs certain functions. But functionalism is to the effect that what matters is not what performs the functions, but that they are performed. So why shouldn't a different kind of experience have performed exactly the same functions as are actually performed by kblue? That would contradict your claim that functions determine the character of your experiences. There would be the same functions but different experiences.

Reply: There's a misunderstanding here. True, functionalism is to the effect that having a particular kind of experience is a matter of certain particular functions being performed. This means there is no more to the experience than the performance of those functions. It follows that there's no possibility of the same functions being performed while I was having a different experience: the experience just is the performance of those functions. The misunderstanding arises from the fact that while on the one hand this experience of kblue is a matter of these particular functions being performed, it is perfectly possible for that experience *itself* to perform functions, such as enabling me to recall similar experiences in the past. To avoid that misunderstanding, we only need to distinguish between any functions the experience may happen to perform and the functions whose performance actually constitutes the experience.

Functions are pretty abstract; but having a toothache, smelling eucalyptus, and tasting Stilton cheese are as real as anything. How can the concrete facts of conscious experience depend on abstract items?

Reply: I don't say that consciousness *is* the functions. It is the *performance* of functions – and what performs them are (in human beings at least) the same concrete processes as are the objects of research in the neurosciences. These immensely complex processes involve the transmission of patterns of activity through multifarious electrochemical connections among billions of neurones. Functionalism provides a way of thinking about what are in fact the workings of the brain, or whatever non-neural processes perform the functions of brains in artefacts or non-terrestrial creatures. To redescribe these brain processes in terms of the functions they perform is not to bring mysterious abstract extras on to the stage. The abstractness of talk about functions springs from the fact that physical details don't matter from the point of view of what it takes for something to be conscious. If the same functions were performed by different physical processes, or by non-physical processes, if that is possible, there would still be consciousness. So, regardless of what performs the functions, having experiences is a matter of absolutely concrete processes. The worry about abstractness is a red herring.

Reflections on functions and relations

Whether or not certain physical items are performing specified functions depends on how those physical items and their activities are related among themselves. As Aristotle pointed out, relations between things often determine both what functions they have and what descriptions apply to them. This piece of stone is a lintel when fixed across the top of a doorway; fixed across the bottom it is a threshold. Its function and description are determined by its position relative to other things. Computers provide more complicated illustrations of the same point. The same circuits may serve to store data, to perform calculations, to transmit commands, and more. And although our brains include a wide range of rather differently specialized kinds of neurones, a single kind of neurone may perform different functions (this is the plasticity of the brain). Recall the example of cochlear implants: the field of neuroprosthetics provides evidence that what matters even for conscious experience is not what physical items perform the necessary functions, but *that* they are performed.

3. Other kinds of consciousness

Although I think the presence of incoming perceptual information is both necessary and sufficient for perceptual consciousness, I have to admit it is not necessary for all kinds of phenomenal consciousness, for example dreaming and daydreaming. (Some philosophers have argued that dreaming is not a matter of having conscious states when asleep; but then, as Descartes remarked, there's hardly any strange view that some philosopher has not held.) They are not cases of perceptual consciousness because perception requires the acquisition of information about external things, while dreaming and reverie don't, even though they have plenty of indirect connections to the outside world. Does that undermine my position?

Those cases show that not all the functions involved in the basic package and presence are necessary for all kinds of consciousness. Further, since the other kinds need not depend on information coming in from the senses, or indeed on any particular relationship between the subject and the outside world, a general functionalist account – one covering non-perceptual as well as perceptual consciousness – had better not require consciousness to depend on the system's external relations (which is a good reason, by the way, for being sceptical about some claims made about the 'extended mind'). Not being in the business of armchair psychology I will not speculate on just what internal processes those other types of consciousness might involve. However, it seems clear that the other types *resemble* conscious perception in many ways; moreover they typically depend at least indirectly on perceptual information which, though not streaming in via the sense organs, is stored up. And there seems to be no particular difficulty over the idea that some stored information forces itself on us much as incoming perceptual information does (see Chapter 9). In daydreaming, of course, we sometimes choose what thoughts and images to evoke – lucid dreaming is another example – but that is consistent with others turning up willy-nilly. I do not have a worked-out functional account of non-perceptual consciousness. However, discussing a pair of (highly artificial) examples will lead to one or two further tentative suggestions.

4. Brains in vats and buckets

The brain in a vat

A thought experiment much used in other contexts (often to illustrate claims about the meanings of words and the contents of thoughts) is a 'brain in a vat': a human brain disconnected from a body but kept alive in a vat and supplied with suitable nutrients, together with a stream of inputs to its nerves in patterns and sequences that, in a normal human being, would have been caused by stimulation from the sense organs. Not being properly connected to a body, this brain has no sense organs. However, it is connected to a science-fictional machine which feeds in a stream of stimulation exactly matching what would have been caused by things in the outside world acting on a normal person. Suppose, then, that by a monstrous coincidence the inputs manufactured by this machine over a certain period of time are exact duplicates of the sensory inputs to my own brain over a similar period, and that at any instant the total state of this brain, which I will call B, exactly matches the state of my own brain at the same instant. Evidently, B would be a counter-example to functionalism if either (a) B would be conscious but the right functions would not be performed; or (b) the right functions would be performed but B would not be conscious.

Would B be conscious? Unscientific sampling among non-philosophers suggests a strong tendency to assume it *would* be conscious. The same seems to be true of philosophers without an axe to grind. On the other hand there is a lot of axe-grinding in this neighbourhood. According to some theories, B would not be conscious at all, in spite of intuitions to the contrary. (Behaviourism is one of these theories, but we have seen good reasons to dismiss it.) The other theories are externalist: they hold that conscious experiences are just a matter of having mental states with certain kinds of contents – states that are *about* certain things – these contents in turn being determined by the subject's external relations (2§7). On any functionalist account the question whether B is conscious boils down to the question whether the right functions are performed in it: B is conscious if and only if the right functions are being performed. But are they?

For the view that the right functions are being performed in B is the fact that it is an exact duplicate of my own brain, where I have to suppose the right functions are performed. Clearly, these will be internal functions: involved in such processes as classifying, storing, and retrieving information. *Against* the view that the right functions are being performed are behaviourism and externalism. Both imply it is a mistake to take only internal functions into account. Before discussing this case further, here is one even more bizarre.

Consciousness in a bucket?

Suppose the random jiggling of atoms in a bucket resulted in their exactly duplicating the successive states of my own brain during one of my periods of normal consciousness. It is counter-intuitive to say there would be consciousness in the bucket. But can I consistently deny it? One thing worth noting is that I could not have denied it if I had maintained that mental states were *identical* with physical states. For since the bucket's states duplicate those of my brain, those identities would have ensured that whatever was true of my brain states – including their being conscious – was true of the bucket states. I have explained why I think the identity theory is mistaken, and I am therefore not committed to its consequences. Still, what should functionalists say about the bucket?

Either (a) the right functions are being performed by the random churnings of the atoms or (b) they are not. On the face of it (a) seems absurd. How could such merely random activity involve the performance of any functions at all? Surely that would have required some purpose to be served, or some intention to be had in mind, or some further objective to be contributed to – all ruled out by randomness. There is also the consideration that it doesn't seem right to think of the contents of the bucket as forming a system, or indeed of having any particular structure. For presumably a system is a kind of structure whose components and workings are in some way coordinated (true even of weather systems). And a structure must surely have certain more or less persisting features, which contribute to determining the course of events. That is certainly the case with brains. Huge numbers of interneuronal connections persist through time, and their character determines how incoming stimuli are processed. In contrast, the randomness of what happens in the bucket seems to imply that the stuff inside it has no particular structure. In the absence of a structure, though, there seems no way for what is going on to be counted as a system in which functions are performed – whether well or ill.

If on the other hand (case (b)) the jiggling atoms are indeed not performing any functions, then my official view has to be that there are no conscious states in the bucket either. But this has its own difficulties. By hypothesis the successive states and configurations of the atoms exactly duplicate those of my own brain, at least for a while. If they are not enough for consciousness by themselves, what can ensure they perform the right functions in my brain? It has to be their relations to other things. But the trouble is this. If their relations among themselves in the bucket are not enough – so that they also require appropriate connections to the outside world – how can it be that what makes the hugely significant difference between consciousness and its absence is nothing more than the connections between the atoms' states and configurations on the one hand, and external things on the other? The idea would have to be that if those external things are absent, as in the bucket case, there is no consciousness; if they are present, there is consciousness. That seems like an appeal to magic.

5. Is the bucket-brain conscious after all?

I am inclined to think the brain in a vat has conscious states, but I am also impressed by the reasons for thinking that the bucket-brain does *not* have conscious states. Yet it is hardly possible to hold both views.

Back to the vat-brain. The 'externalist' theories according to which it is not conscious maintain that consciousness depends either on dispositions to overt behaviour, or else on what the system's states are *about*: on their contents. Since behaviourism has been ruled out, this latter view (representationalism or intentionalism: 12§5) is the main source of resistance to the idea that the vat-brain is conscious. But these externalists cannot deny that plenty of unconscious mental states have content. (You don't have to be a Freudian to agree there are unconscious mental states: your current beliefs about historical or geographical facts are not all conscious now.) So they owe an explanation of what it is about the contents of certain mental states that ensures they are conscious while other contentful mental states are not. The most popular explanations appeal to 'higher-order' thoughts: thoughts about thoughts. In view of the criticisms mentioned when this view was considered earlier (12§5), I will assume that those explanations are wrong, and that therefore the brain in a vat is conscious after all.

If that is right, I am forced to hold that performance of the right kinds of internal functions in a system is by itself enough to make it conscious-in-general, and also enough to fix what it is like for it to be in its various conscious states. So what must I say about the bucket-brain?

Recall how it was defined. The random jiggling of the atoms in the bucket was said to result in their exactly duplicating the successive states of my own brain during one of my periods of normal consciousness. I originally suggested that this randomness bars the bucket's contents from having any structure, disqualifying them from performing any functions. But on further reflection I have come to realize that that reasoning was unsound. If the bucket's contents really do duplicate my brain's successive states for the period in question, then they are indistinguishable from my own brain for that period, so that no difference between the two would be detectable even by experts. And (taking for granted that the laws of nature do not change) once the random play of the atoms has led to the formation of a totally brain-like structure, this structure will continue to exist just as a normal brain would have done – although, of course, unlike the vat-brain, we cannot expect the inputs to its afferent nerves to continue to resemble those that would have been caused by the impact of a real external world. So, over the period in question, the bucket-brain has the same structure as my brain, regardless of whether this state of affairs has come about randomly or non-randomly. If that is right, whatever internal functions my brain performs are performed by the brain in the bucket. In spite of my initial reluctance to accept that brain-like internal functions are being performed in the bucket, therefore, that is what is going on. It looks as if functionalists must say that both vat-brain and bucket-brain have conscious states, far-fetched though these cases are.

Can we stick with that conclusion? Are there any special difficulties over the suggestion that performance of the right kinds of *internal* functions is not only necessary but also sufficient for both consciousness-in-general and particular conscious states – even in isolated systems without connections to sense organs and controllable bodies?

6. A conscious lump of matter?

At this point it will be useful to revise the vat-brain story. I said it was a human brain disconnected from a body but kept alive in a vat. To make it more closely comparable to the bucket-brain, I now redefine it as having always been in its vat. This new story has two further important features. One is that the stream of inputs to this *ab initio* vat-brain's receptors is not caused by machinery simulating the impacts of the real external world on sense-receptors, but fictitious. The inputs are what they would have been if the individual had been a normal person born in a real world which, however, did not resemble our world. The other important revision is that the nature of the inputs is the result of chance events (fortunately there is no need to specify them). To pile cosmic fluke upon cosmic fluke, this *ab initio* vat-brain did not develop naturally; it started its existence as a chance conglomeration of atoms which happened to be exactly like the brain of a newborn infant. You may find this version of the story too repellently improbable to think about, but, as usual, our problems are theoretical. And as the suggestions of theoretical physicists show, it can be instructive to consider very outlandish possibilities.

Since the *ab initio* vat-brain has no sense organs, it lacks the links with the outside world that would have allowed us to say that some of its internal processes performed the functions of (for example) memory-storage, or that others were involved in shape recognition. It cannot recall conversations with its next-door neighbours, for example, because no such beings exist: its apparent memories are pseudo-memories caused by random processes. Further, we cannot ascribe functions to its internal processes on the basis of what it was designed or intended to do because, being a product of chance events, it has no designers. Nor can it have functions determined by evolution because it didn't evolve.

I said earlier that the functions in question would be performed by processes of classifying, storing, and retrieving *information*. Does that give us a way out of the present difficulties? Hardly. Given we are now thinking about an *ab initio* vat-brain with no connections to the outside world, could it really be described as processing information? Compare computers. What goes on inside them is often described as processing information because normally (at any rate) much of the data stored in them will represent the results of human activity: about such things as the symptoms of diseases, or facts about legal proceedings, or train timetables. But the same is not true of what goes on inside the vat-brain or the bucket-brain – because they have no connections to external sources of information.

If these systems cannot properly be described as either performing functions or processing information, is there any hope for a functional explanation of their consciousness?

I suggest the difficulties we have just been considering are only verbal. It might indeed be inappropriate to describe what goes on in vat- and bucket-brains as processing information, or to say that this processing involves the performance of functions. Yet the view I am inclined to take can still, I suggest, reasonably be described as functionalist. This is because it makes the presence or absence of consciousness depend on relations between the components of the system – how they interact and work together, on their *intrasystemic* relations – and not on what they are made of. A comparison may be helpful. Over two hundred years ago William Paley, expounding the 'design argument' for the existence of God, imagines that in crossing a heath he finds a watch on the ground. He contrasts this with finding a stone:

> when we come to inspect the watch, we perceive (what we could not discover in the stone) that its several parts are framed and put together for a purpose. (Paley 1802: 1–2)

He reasonably supposes that such a thing could not have come into existence by chance. But even if it had, we should still be able to discover how it worked. We should be able to use it as a watch and find it natural to say its components performed functions: these are the functions they *would* have performed if it had been designed and constructed for the purpose of performing them.

I suggest the same is true of the vat- and bucket-brains. If either of them were to be connected via suitable nerves to a normally working human body, all the right functions would be performed, for there would be no significant difference between the resulting individual and a normal person. Both internally and in their relations with the outside world they would be just like us: normally conscious. And just as it is natural to say the components of the chance-produced watch perform the functions of a normal watch, so it is natural to say the components of vat- and bucket-brains perform certain functions (not *all* the functions performed in our own brains, of course, because some, like those involved in perception and memory, depend on external connections).

It is natural to put things in those terms but not compulsory. If there are better ways to refer to the workings of these systems, I would like to know what they are, although I need to be persuaded that it matters. What really does matter, I suggest, is that the presence or absence of consciousness depends on how the system's components interact and work together: on their intrasystemic relationships and not on what they are made of. If its internal workings mirror those processes in a normal human being which provide for the presence of incoming perceptual information, then, I suggest, that system is conscious. That is true even if, as with the vat- and bucket-brains, no perceptual information is actually coming in because there are no sensory links with the outside world. What matters is that there are those kinds of intrasystemic processing. So I am defending what might be called an 'internal-functionalist' position.

Finally, the kinds of intrasystemic processing that matter don't depend on any particular kinds of physical (or non-physical) components. The system might consist entirely of bits of metal and plastic, and its low-level elements might be electronic, provided its internal processing satisfies the conditions mentioned. In short, any object in which the right internal functions are performed is conscious.

To end the book I will recall three key points.

7. Three key points

1. *Zombies and transposed qualia are logically impossible.* The steel rails of our intuitions about zombies and transposed qualia seem to compel us to think that conscious experience is so special it could not be provided for by anything physical or functional. But I have argued that if those were genuine logical possibilities, epiphenomenalism might be true – when in fact it involves a contradiction and therefore couldn't possibly be true. Too easily we form a mistaken conception of the nature of consciousness: we tend to think of it as a special stuff or property unlike anything physical or functional. The argument in Chapter 8 shows, I think, that there could be no such special stuff or property. Conscious perceptual experience is inextricably linked with abilities to do such things as notice, attend to, and remember experiences. It cannot – logically – be separated from those abilities, while speculations about zombies and transposed qualia depend on assuming they could be separated.

2. *We must all be functionalists.* By demolishing the intuitions, the anti-zombie argument removes what has long been seen as a major problem for physicalism and functionalism. And by bringing out the fact that consciousness depends on a number of different information-processing capacities, that argument hints at what it really is. Instead of thinking of consciousness as a special stuff or property, we think of it as the performance of functions, notably those involved in the basic package plus the presence of incoming perceptual information – and what is essential about those functions could in principle occur even in an isolated lump of matter. In fact there is no alternative to a functional explanation of perceptual consciousness.

However, you might still be bemused by the idea that consciousness is nothing but the performance of functions. But of course that notion still strikes many people as absurd. In the Preface I quoted John Searle's remark that those tempted to functionalism need help rather than refutation. I think that on the contrary it is those who reject functionalism who need refutation. They also need help, though, and in this book I have been trying to make functionalism more appealing than the assumptions and intuitions which feed misconceptions about it.

One thing that may be getting in the way is the word 'function' itself; perhaps it is too abstract. But, as I emphasized a little way back, the idea is not that having an experience is itself a function, but that it is the *performance* of certain functions: in human beings by hugely complex neural processes; in artefacts by equally complex electronic or other processes. After all, the neurosciences have been discovering more and more about how the brain provides for the different functions involved in perception (analyses of coloured light, classifications, representations of various kinds, memory, and a vast amount else). And no one suggests the brain is incapable of performing the necessary functions.

3. Redescription. So here is the third key point.'First-person' descriptions of experiences in terms of what it is like to have them are redescriptions – what I am calling 'pure' redescriptions – of whatever processes are performing the relevant functions. Typically, as noted earlier, we don't have any idea what those processes are, or even what functions they are performing. It is not surprising, then, that the truly special nature of consciousness, together with the independence of phenomenal concepts from physical and functional concepts, has made it easy to suppose that it must involve special non-physical properties. Yet performance of the relevant functions is redescribable in third-person terms, and the redescriptions are pure, since nothing else is involved.

The situation might have been illuminated by a neat analogy. Unfortunately no other phenomenon is much like consciousness. Perhaps, though, a look at the windows of Notre Dame in Paris will help. Some of the stained glass in that cathedral is of an extraordinary slightly purplish blue. Viewed from outside, you just see these dark-looking windows; to appreciate their intense colours you have to be inside. The external view might be said to correspond to learning in third-person terms about the functions being performed when we actually see the windows. Being inside and viewing the windows with light streaming through would correspond to actually having the experience constituted by the performance of the functions in question. The comparison has some appeal because, when we are inside the building and see the vivid colours of those windows, we know that nothing else is involved than what we were already acquainted with: light coming through the windows. Similarly, we know – according to the approach I favour – that nothing else is involved in conscious experience than the neural or other processes which constitute performance of the relevant functions.

Unfortunately this analogy echoes the discredited model of a homunculus viewing an internal cinema show, when there can be no such show and no such viewer (9§5). Someone has to see the windows – and that someone is our little old friend the homunculus. Yet it remains true and instructive that the cathedral glass viewed from outside (compare third-person descriptions) is the same as the glass viewed from inside (compare first-person descriptions). So the analogy still has a use: it helps to make vivid the crucial point that performance of the functions involved in presence is *the very same thing* as having conscious perceptual experiences.

8. Bottom lines

What is intelligence? The basic package. What is consciousness? The basic package plus presence. Could robots be genuinely intelligent? Yes. Could they be genuinely conscious? We have seen no reason why not. Could there be zombies? No, they are logically impossible.

References

Armstrong, D. M. (1968), *A Materialist Theory of the Mind*, London: Routledge and Kegan Paul.
Berkeley, G. (1710), *A Treatise concerning the Principles of Human Knowledge*. (Many editions available.)
Block, N. (1981), 'Psychologism and Behaviourism', *Philosophical Review* 90: 5–43.
Chalmers, D. J. (1996), *The Conscious Mind: In Search of a Fundamental Theory*, New York and Oxford: Oxford University Press.
Damasio, A. (1999), *The Feeling of What Happens: Body, Emotion, and the Making of Consciousness*, London: Heinemann.
Dennett, D. C. (1969), *Content and Consciousness*, London: Routledge and Kegan Paul.
Dennett, D. C. (1987), *The Intentional Stance*, Cambridge, MA: MIT Press.
Dennett, D. C. (1991), *Consciousness Explained*, Boston, MA: Little, Brown.
Descartes, R. (1637), *Discourse on Method*. (Many editions available.)
Descartes, R. (1641), *Meditations*. (Many editions available.)
Dretske, F. (1995), *Naturalizing the Mind*, Cambridge,,MA , and London, England: The MIT Press.
Fodor, J. A. (1975), *The Language of Thought*, New York: Thomas Y. Crowell.
Fodor, J. A. (1992), 'The Big Idea: Can There Be a Science of Mind?', *Times Literary Supplement*, 3 July.
Hillyard, P. D. (1996), *Ticks of North-West Europe*, Shrewsbury: Field Studies Council for the Linnaean Society of London and the Estuarine and Coastal Sciences Association.
Hofstadter, D. R. and D. C. Dennett (eds.) (1981), *The Mind's I*, New York: Basic Books.
Jackson, F. (1982), 'Epiphenomenal Qualia', *Philosophical Quarterly* 32: 127–36.
Kirk, R. (1974a), 'Sentience and Behaviour', *Mind* 83: 43–60.
Kirk, R. (1974b), 'Zombies v. Materialists', *Proceedings of the Aristotelian Society, Supplementary Volume* 48: 135–52.
Kirk, R. ([2005] 2008), *Zombies and Consciousness*, Oxford: Oxford University Press.
Kirk, R. (2008), 'The Inconceivability of Zombies', *Philosophical Studies* 139: 73–810.

Kripke, S. (1972), *Naming and Necessity*, Oxford: Blackwell.
Levine, J. (1993), 'On Leaving Out What It's Like', in *Consciousness: Psychological and Philosophical Essays*, M. Davies and G. Humphreys (eds.). Oxford: Blackwell, 121–136.
Levine, J. (2001), *Purple Haze: The Puzzle of Consciousness*, Oxford and New York: Oxford University Press.
Locke, J. ([1689] 1975), *An Essay Concerning Human Understanding*, P. H. Nidditch (ed.), Oxford: Clarendon Press.
Lucas, J. R. (1961), 'Minds, Machines and Gödel', *Philosophy* 36: 112–27. Reprinted in A. R. Anderson (ed.), *Minds and Machines*, Englewood Cliffs, NJ: Prentice-Hall (1964), 43–59, to which page references apply.
Nagel, T. (1974), 'What Is It Like to Be a Bat?', *Philosophical Review* 83: 435–50. Reprinted in his *Mortal Questions*, Cambridge, England: Cambridge University Press, 1979.
Paley, W. (1802), *Natural Theology, or Evidences of the Existence and Attributes of the Deity Collected from the Appearances of Nature*, London: R. Faulder and Son.
Putnam, H. (1975), 'The Meaning of "Meaning"', in *Mind, Language and Reality: Philosophical Papers* vol. ii, 215–71, Cambridge: Cambridge University Press.
Rosenthal, D. M. (1986), 'Two Concepts of Consciousness', *Philosophical Studies* 49: 329–59.
Ryle, G. (1949), *The Concept of Mind*, London: Hutchinson.
Sartre, J.-P. (1958), *Being and Nothingness: A Phenomenological Essay on Ontology*, trans. Hazel E. Barnes, London: Methuen. Originally published as *L'Être et le Néant*, Paris: Gallimard (1943).
Searle, J. R. (1980), 'Minds, Brains, and Programs', *Behavioral and Brain Sciences* 3: 417–24, reprinted in Hofstadter and Dennett (eds.) 353–73, to which page references apply.
Searle, J. R. (1992), *The Rediscovery of the Mind*, Cambridge, MA, and London: MIT Press.
Smart, J. J. C. (1959), 'Sensations and Brain Processes', *Philosophical Review* 68: 141–56, reprinted, revised, in *The Philosophy of Mind ...* to which page references apply.
Stout, G. F. (1931), *Mind and Matter*, Cambridge: Cambridge University Press.
Turing, A. M. (1950), 'Computing Machinery and Intelligence', *Mind* 59. Reprinted in Hofstadter and Dennett (eds.), 53–68, to which page references apply.
Tye, M. (1995), *Ten Problems of Consciousness: A Representational Theory of the Phenomenal Mind*, Cambridge, MA, and London: MIT Press.
Weiskrantz, L. (1986), *Blindsight: A Case Study and Implications*, Oxford: Clarendon Press.
Wittgenstein, L. (1953), *Philosophical Investigations*, trans. G. E. M. Anscombe, Oxford: Blackwell.

Index

a priori and *a posteriori* (empirical) deduction 153, 154
a priori and logical claims separated 155–6
absent-minded driving 3
abstract machines 23–5
analyticity 136
arguments 7
Armstrong, D. M. 164
artificial intelligence (AI) 21–2, 30
atomic theory 58–9
automata 1, 11, 25
  'conscious' 64–5

Babbage, C. 23
basic package 93–115
  defined in terms of capacities 99
  interdependence of its capacities 103–5
  necessary and sufficient for intelligence 109
  not by itself sufficient for consciousness 113–15
  with presence, sufficient for consciousness 117–23
  and robots 109, 110–13
bats 6
behaviourism 2, 9–17
  psychological 2
  reasons to reject 42, 45
  see also Giant; perfect actor
belief 107
  as dispositional 12–13

Berkeley, G. 54–5
blindsight 4, 113–14, 158–9
Block, N. 37, 38
  neo-Turing conception of intelligence 38, 40
Block's machines 37–41
  programming 39–40
  satisfy the Neo-Turing conception 41
brain in a vat 174
  ab initio 178–9
  consciousness in 174–5, 180
bucket-brain 175–8, 180

Cartesian intuitions 138, 151, 152, 167
Cartesian Theatre 82, 121–2
Chalmers, D. 166
  arguments for possibility of zombies 70–4
chess-playing programs 37–8
  brute force 37
  Deep Blue 37
  sophisticated 37–8
Chinese Room argument 30, 31–7
  Robot Reply 35–7
    Searle's reply 36–7
  Systems Reply 32–5
    Searle's reply 33–5
cochlear implants 29, 172
computers 19–22
  basic features 20–2

how they work 20–1
machine code 21
memory 20
and other machines 19–25
programs 20
as Turing machines 45
see also machines; robots; Turing machines
computers-for-neurones system 148–50
  not a counter-example to functionalism 149–50
conceivability 49–50, 152
concept possession and conceptualization 105–9
  and language 105–7
  not all-or-nothing 107–9
concepts 105–7
  of colour 153–4
  phenomenal 130
  psychological 119–20
  viewpoint-neutral (third-person) 130–1, 154
  viewpoint-relative (first-person) 130–1, 134, 153, 154, 157, 159, 182–3
consciousness *passim*
  consciousness-in-general 140
  depends solely on intrasystemic processes 180
  evolution of 4, 115, 158
  in an isolated lump of matter 178–80
  meanings of the word 5–6
  mistaken conceptions 58, 82, 126, 157–8
  not explained by dualism, idealism, dual aspect theory, panpsychism or physicalism 62
  our knowledge of it 93–4
  perceptual 94, 117–23

  as performance of functions 140–3
  phenomenal 6, 94, 130, 154
  rival accounts 159–65 (*see also* Sartre; sense of self; representationalism; higher-order thought; inner perception; Wittgenstein)
  special access to conscious states 156
  special knowledge of our own conscious states 157
  its use 4, 115
  what matters 180
  see also presence; what it is like
control of behaviour 98–9
  See also basic package
Crick, F. 64

Damasio, A. 161
Dan 68–9
decider 98–9
  see also basic package
Dennett, D. 7, 10, 102
Descartes, R. 10, 11, 16–17
  on animals as automata 48
  argument for distinctness of body and soul 49–50
  on the body as machine 47, 48
  on conceivability 49–50
  humans are not machines 47–50
  malicious demon 16–17, 54
determinism 29
dispositions 9, 10–11, 16
  and beliefs and wants 12–13
  in computers 11
  and sensations 12
dual aspect theory 55–6
dualism 2, 47–54, 89
  Cartesian 47–50
    objections 50–2
  see also epiphenomenalism; interactionism; parallelism

epiphenomenalism 52–3, 76–7
  big problem 77–8
  counter-examples 80–2, 83
  not logically possible 82, 86
  wrong conception of consciousness 82
everyday psychology *see* psychology
evolution 4, 115, 158
experiences 6, 53, 78
  see also consciousness
explanatory gap 151, 152
  epistemic gap 166
  no ontological gap 166
  ruled out by functionalism 159
  two crucial facts: special access, special knowledge 156–7
extended mind 162
externalism 16–17

first-person concepts *see* concepts
Fodor, J. 6
folk psychology *see* psychology
free will 29, 100–1
function 1, 127–9
  characterizable in terms of basic package and presence 128–9
  dependent on our interests 128
  as something done 127–8
functional description 26, 132
  logically sufficient for mental descriptions 138
functional isomorphism 126, 140
functionalism 5, 61–2, 125–38
  and causation 125–7
  compulsory 126, 139–47, 181
  does not require mental states to be functionally definable 130
  functional facts fix what it is like 166, 169

functional truths entail phenomenal truths logically but not *a priori* 155–6
must be 'deep' rather than 'surface' 129–30, 144
objections discussed 130–5, 139–40, 147–8, 148–50, 151–2, 153–5, 158–9, 170–2

ghost in the machine 10
Giant 14, 111, 129
  doesn't have thoughts or feelings 14–15
Gödel's theorem used against mechanism 42–5

Hephaistos 1, 25
higher-order thought 164–5
Hobbes, T. 59
homunculus 122, 183
Huxley, T. H. 53

idealism 54–5, 87
identity theory 60–2
  contrasted with functionalism 125–6
  not sufficient for physicalism 60–1
  property objection 61
  redundant if functionalism true 61–2
information 95, 97–8, 101–3
  acquisition (learning) 95, 97, 101–3
  in the brain 102
  called up 115, 117, 118
  in computers 20, 101
  requires the basic package 97–9, 167
  requires control of behaviour 98
  requires information the system can use 97–8
  for the system 103

informational contact 78–9
  ruled out by epiphenomenalism 82
inner perception 163–4
intentionality 27, 35, 46
interactionism 47, 51, 89–90
intuition pump 7, 69, 72
inverted spectrum *see* transposed qualia
inverters 146–8

Jackson, F. 73, 153–4
  Mary 73, 154, 166

Kripke, S. 159

language and meaning 137
  meanings 'just ain't in the head' (Putnam) 17
learning *see* information
levels of description and explanation 26, 44, 102
Levine, J. 152
Locke, J. 71, 94, 95, 154, 163, 167
logical entailment (implication) 137–8, 156

machine table 24, 25, 45, 110
  slot machine example 24
machine table robot 110–13
  lacks the basic package 111–12, 129
  as objection to behaviourism 112
machines 19–26
  abstract 23, 25, 31
  can they think? 30
  mechanical and electronic 19, 22–3
  Watson and Emilia 25
  see also robots; Turing machines
materialism *see* physicalism
McGinn, C. 6

mechanism 42–5, 48
mental concepts not functional 130–2
mind-body problem 6
monitoring behaviour *see* control of behaviour
multiple realizability 60, 62

Nagel, T. 6
Neo-Turing conception of intelligence 38
Notre Dame analogy 182–3

occasionalism 51
oysters 94–5, 108, 167

Paley, W. 179
panpsychism 56–8, 87–9
parallelism 53–4
perception 94
  as acquisition of information 95, 167
  and pure reflex systems 9
  requires control of behaviour 98–9
  requires learning 95
  subliminal 3–4
  unconscious 3
  see also basic package; learning; rabbitoid
perceptual consciousness *see* consciousness
perfect actor 13–14
  fails as counterexample to behaviourism 14
phenomenal truths, concepts, properties 57, 130, 154, 169
  logically but not *a priori* entailed by functional truths 155
  phenomenal truths as redescriptions of the performance of functions 137–8
  why viewpoint-relative 156–7

physical and functional facts
   do not entail phenomenal truths *a priori* 154
   yet do entail them logically 156
physicalism (materialism) 3, 41, 58–62, 87, 91
   see also identity theory
point of view 133, 134, 157, 167
   explained by functionalism 157, 159
possibility natural and logical 67
pre-established harmony 51
presence (of perceptual information) 117–23
   defined 118
   involves a whole complex of processes 122–3
   involves no dilemma 121–3
   necessary and sufficient for perceptual consciousness 150
   scientifically respectable notion 119–21
programs *see* computers
psychology 41–2, 119
   concepts 119
   moderate realism 41–2, 100
   vagueness 119
psycho-physical identity theory *see* identity theory
pure reflex systems 48, 96–8
   cannot learn 97–8
   contrasted with many computer programs 11
Putnam, H. 17, 75

qualia 75
   inert according to epiphenomenalism 84–5

rabbitoid 114–15, 117, 123
   as a decider 114

realism of everyday psychology 100, 129
redescription 44–5, 102, 118, 121, 156
   definition of 'pure' redescription 135–8
representationalism 163
robotics 21–2
robots 1–2, 25, 27–46
   Amelia and Watson 25
   and the basic package 109, 110–13
   and free will 29
   golden servants 1, 25, 28, 113, 115
   and intelligence 27–46, 109
   misguided objections 28–30
   and souls 29–30
   see also artificial intelligence; Chinese Room argument; machine table robot; Turing test; Robot Reply, Chinese Room argument
Rosenthal, D. 165
Rotterdam sea defences 100–1
Russellian monism 56, 57
Ryle, G. 10, 16, 164

Sartre J.-P. 161
science 16, 56, 63, 170
Searle, J. viii, 5, 30–7, 93
   see also Chinese Room argument
sense of self 161–2
stimulus-response (S-R) systems *see* pure reflex systems
Stout, G. F. 65, 66
Systems Reply *see* Chinese Room argument

third-person concepts *see* concepts
thought experiments 7, 15–16, 69–73, 174–80

transparency 151, 152
  and consciousness 152
  why impossible for a functional account of consciousness 156–7
  see also explanatory gap
transposed qualia 71–2, 143–7
  as logically impossible 144–7, 181
  possible given functional isomorphism with physical difference 147–8
Turing, A. 30
Turing machines 23–35, 30, 45
  inputs, outputs, internal states 23, 24
  slot machine as illustration 24
Turing test 30, 37
  inadequate as a test of intelligence 42

viewpoint-neutral, viewpoint-relative *see* concepts

what it is like 6, 7, 57, 61, 73, 74, 141, 143, 144, 153, 154, 168, 177, 182
  fixed by functional facts 166, 169
Wittgenstein, L. 160

zombies 63–91
  arguments against their possibility 75–91
  arguments for their possibility 64–74
  defined 65
  and epiphenomenalism 64–5
  folkloric 65
  functional 126
  idea based on mistaken assumptions 76
  are logically impossible 115, 181
  are naturally impossible 67
zombism 76
  requires possibility of epiphenomenalistic world 86–90
Zulliver 69–70

www.ingramcontent.com/pod-product-compliance
Lightning Source LLC
Chambersburg PA
CBHW050139240426
43673CB00043B/1734